MOORE'S IMAGING DICTIONARY

MOORE'S IMAGING DICTIONARY

The Official Dictionary of Electronic Document and Image Processing

Andy Moore

Includes
OCR/ICR Compression Digital Video Document Imaging
Photo Manipulation Optical Storage

A Telecom Library, Inc. Book
Published by Telecom Library, Inc.
Copyright © 1993 by Andy Moore
All rights reserved under International and Pan-American
Copyright Conventions, including the right to reproduce
this book or portions thereof in any form whatsoever.
Published in the United States by Telecom Library, Inc.,
New York.

ISBN 0-936648-37-6

Manufactured in the United States of America

August 1993
Cover Designed by Saul Roldan.
Printed at BookCrafters

Moore's Imaging Dictionary

This dictionary wouldn't have been possible without the steady support and fierce dedication of the staff of Imaging Magazine: Harry Newton, Kim Ann Zimmermann, Marta Neilson, Jules Gilder, Mike Popovic, Aaron Brenner and Lee Mantelman.

I've mined many sources in pulling together the Imaging Dictionary. Among the best in their fields: Delphi Consulting's (Boston, MA) guide to text retrieval, document management and workflow; Xplor International's (Palos Verdes Estates, CA) guide to electronic document systems; Nisca's (Carrollton, TX) guide to hand scanning; Digital F/X (Mountain View, CA) video glossary; Hewlett-Packard's (Cupertino, CA) optical storage primers; and Harry Newton's (New York, NY) Telecom Dictionary (also published by Telecom Library Inc.). Thanks also to Gwen Hoover from Calera Recognition Systems (Sunnyvale, CA).

Moore's Imaging Dictionary is still growing. I will update and expand it every day. Send me new words, terms and acronyms. I'll put them in the next edition. We'll both be famous! Send additions to:

Moore's Imaging Dictionary
12 West 21 Street
New York, NY 10010
Fax: 212-691-1191. Computer: 212-989-4675. MCI Mail: 529-3434.

Typography and book design by Jennifer Cooper-Farrow.

Jacket design by Saúl Roldán.

Dedicated to Maggie and Sleepy

Moore's Imaging Dictionary

A The first letter of the alphabet. In ASCII, uppercase "A" is represented as hexadecimal 41; a lowercase "a" is hexadecimal 61. In EBCDIC, an uppercase "A" is hexadecimal C1; a lowercase "a" is hexadecimal 81.

A A noise word. That is, a word that full-text retrieval systems ignore because no one will ever want all the documents with the word "A" in them. Will they?

AA Authors Alterations. Corrections to text after it has already been committed to typesetting. Try to avoid them. It's a printer's favorite line item, i.e. they add lots to your bill.

A/B ROLL A video editing technique where you have a scene or sound on two different reels and play (roll) them both together to perform a transition or special effect using both scenes or sounds at once. A term from the film world, referring to editing with two separate — "A" and "B" — rolls of film. When editing film, the audio track is also edited separately. Two sources or tapes are necessary when dissolving from one scene to another (in video, the audio portion is usually an integral part of one or both of the videotapes.) This process requires two source VTRs that are both frame-accurate, and two time-based correctors.

AC Alternating current. Typically refers to the 120 volt electricity delivered by your local power utility to the three-pin power outlet in your wall. Called "alternating current" because the polarity of the current alternates between plus and minus 60 times a second. The other form of electricity is DC, or direct current, in which the polarity of the current stays constant. Direct current, for example, is what comes from batteries. Outside North America, electricity typically alternates at 50 times a second — which is neither better nor worse, just different.

In North America, "standard" 120 volt AC may actually come out of the wall at 110 volts, 115 volts, 117 volts or 125 volts. NYC's utility, Con Edison, says they are obliged to deliver 120 volts plus or minus 10%. Your outlet may deliver anywhere from 118 volts to 132 volts.

Active video The portion of the video signal that holds the picture. The idea is to keep important things away from the edge. The "action safe" area is 5% in from all four sides of the active video. The "title safe" area is 10% in from all four sides.

A/D converter Analog to Digital Converter. Generic term for any device that changes continuous wave (analog) activity to binary (digital) code, or vice versa. Modems and codecs are A/D converters.

Abend Abnormal END or ABortive END. Also called system crash. Almost always bad news. Usually caused by input or data presented to a computer which is beyond its ability to cope. If an abend happens in a single-task program (like MS-DOS), the machine will cease to take input ("lock up") and must be restarted ("re-booted"). Multitasking operating systems (like UNIX) allow other programs to continue running while only stopping the one causing trouble.

Ablate To remove. Used in reference to the formation of laser-readable "pits" in the recording layer of optical discs.

Accelerator board A printed circuit board added to a PC to increase its performance.

Accept rate The percentage of characters an OCR recognizes (either correctly or incorrectly) from the total number of characters in the document.

Access arm The device in a disk or disc drive that holds the read/write heads and moves them into position.

Access method The technique or the program code in the operating system that provides input/output services. It defines where a group of data will be stored on a medium (see below). By including the access method in the basic operating system, computer makers have made the programmer's job much simpler.

In the case of tape drives, the access method is straightforward — a block of data is placed sequentially after the last one. Disk drives present a different challenge.

Disk drives do not (usually) place data in sequential tracks. The access method software places a block of data in an available empty space, and creates an index (called a File Allocation Table, or FAT) that notates where the block of data can be found for retrieval later.

Access time How long it takes a magnetic or optical drive to find data. It is the sum of the average latency and average seek time. In optical drives, seek time is often expressed in "average seek time 1/3 stoke." This means the average time it takes the drive to find a file within 1/3 of the disc's radius. Since optical discs hold 325 megabytes per side, 1/3 stroke is the time it takes to seek from about 107.25 megs. It is NOT 1/3 the time it takes for a full stroke. The two are not directly proportional, and results in misleading access times.

Accuracy A means of judging OCR. The extent to which document is recognized with no errors. Measured as a percentage — 99.6% accuracy on decent-quality documents is the minimum I'd accept.

Acetate-base film A film substrate used in microform production. Considered a safety film (i.e. meets ANSI safety standards.)

ACF Advanced Communication Function. A series of IBM software products that controls networking tasks.

ACF/NCP Advanced Communication Function for the Network Control Program. An IBM term.

ACF/VTAM Advanced Communication Function/Virtual Telecommunications Access Method. An IBM term.

Achromatic Means having no color. Fancy way of saying black-and-white. Many local newspapers are achromatic.

Acoustics The branch of science pertaining to the transmission of sound. The qualities of an enclosed space describing how sound is transmitted, e.g. its clarity.

ACK A data communications term. A control character that means a "positive acknowledgment," or "that last block of data was received correctly, OK to send another."

Acquisition A pompous word for scanning. "Image acquisition is achieved by means of a optical digitizing unit" is how military writers would say "We scan paper."

Acrobat Adobe Systems' software that lets you share desktop published documents with other people, even if they don't have the original application or fonts that created the documents.

Actinic The chemical change violet or ultraviolet light produces in certain photosensitive film.

Actionmedia DVI Technology's product family, introduced in 1990 and consisting of single-board delivery and single board capture capability for AT or Micro Channel architecture buses. Introduced with IBM.

Active display area The part of an image that falls inside the borders of the display screen, i.e., the part you can see.

Active matrix LCD LCD display technique that uses a transistor for each monochrome or each red, green and blue dot. It provides sharp contrast, speeds screen refresh and eliminates loss of cursor (submarining) on standard LCD screens.

Active video area The area of a TV monitor excluding the border where clips, special effects and edited video can be viewed. For an NTSC system, the active video area is 624 pixels wide by 464 pixels high.

Active vocabulary In voice recognition, a group of words which a recognizer has been trained to understand and is attempting to understand at a given time. It is a subset of the total vocabulary of the recognizer.

Actuator Mechanism that moves an object; for example, the device that selects a laser disk or CD in a jukebox, or the read/write head on a disk drive.

Adapter 1. A device that links two devices together in a system. Sometimes implies the two were somehow physically incompatible, and an adapter was required to compensate. 2. A device used to control a piece of peripheral hardware — from a joystick to a disc drive.

Adaptive Can handle wide variations in the characteristics of its input, and still do its job. I.e., is flexible.

Adaptive compression Data compression software that continuously analyzes and compensates its algorithm (technique), depending on the type and content of the data and the storage medium.

Adaptive Differential Pulse Code Modulation ADPCM. A digitizing standard that allows a voice conversation to be carried within a 32Kbps digital channel.

ADB Apple Desktop Bus. The interface on a Mac where non-peripheral devices, such as the keyboard, attaches. A Mac keyboard or mouse is called an ADB device. Contrast with peripherals, which attach through the SCSI interface.

ADCCP Advanced Data Communications Control Procedure. A bit-oriented communications protocol standardized by ANSI.

Address 1. Disks and other storage devices have numbers that identify locations by sector and by byte, like addresses on a city street. Retrieval software searches for the address assigned to the desired information in order to locate it. See index.

2. Used as a verb, it means what a computer can access: "This Mac can address 5 megs of RAM."

ADF Adapter Description File. The text file that is associated with a Micro Channel peripheral that describes the resources it needs to operate.

Address register A memory device that holds the location of memory, peripherals or other physical add-ons.

Address space 1. The amount of system locations available to a program. 2. The virtual storage assigned to each program currently running in a system.

Addressability The ability to place information at a certain chosen area in an image. Also, the number of dots per inch that a printer can place on the page. Expressed as dpi: "a 300-dpi thermal wax printer."

Addressable capacity The number of locations on an image that are addressable. To calculate, multiply the addressable vertical positions (row) by the addressable horizontal positions (column). Think of a matrix of dots, eight across by 16 down. The addressable capacity of the matrix is 128.

Additive color Color produced by the combination of red, green and blue, called the additive primaries. It is called additive, because the addition of one color beam of light produces additional light. White light is the combination of all colors of light. Black is the absence of light. See Subtractive color.

Address The location of specific data in a memory device.

Adobe Systems The company which produces PostScript, PhotoShop and Acrobat. Adobe Systems got its name from the creek than ran past its founder's home in Los Altos, California. See also Adobe Type Manager.

Adobe Type Manager Software from Adobe Systems, Inc., Mountain View, CA, for Macintoshes that eliminates jagged edges on screen fonts and allows inexpensive (non-Apple) laser printers to reproduce PostScript fonts accurately and cleanly.

ADP Automatic Data Processing.

ADPCM Adaptive Differential Pulse Code Modulation. A digitizing standard that allows a voice conversation to be carried within a 32Kbps digital channel.

ADR Automatic Dialog Replacement. In video editing, the process of replacing the dialog recorded on-location with dialog recorded in the studio. Also called "looping."

ADSTAR Automated Document Storage And Retrieval. Generic term for systems that identify, select and display images that have been previously electronically stored.

AFIPS American Federation of Information Processing Societies.

AFP AppleTalk Filing Protocol. The protocol that non-Apple networks need to use in order to access data in an AppleTalk server

Agate Type that is 5 1/2 points. Used in newspaper classified ads. There are 14 agate lines to a vertical inch.

AGC Automatic Gain Control. An electronic circuit in tape recorders, speakerphones and other voice devices which controls the volume of the recoded sound. Not a brilliant idea, since it attempts to produce a constant volume level of everything it hears: the volume of your voice, plus the static and/or general room noise which you do not want amplified.

Aggressive recognition An OCR that uses rules and parameters that increase the chance of a mistake. A broader OCR. Results in fewer rejects, but more mistakes.

AI See Artificial Intelligence.

AIIM Association for Information and Image Management. Trade association and professional society for the micrographics, optical disc and electronic image management markets. 1100 Wayne Ave., Suite 1100, Silver Spring, MD 20910. 301-587-8202.

Air The unused space on a page, also called white space.

Airbrush A fine-mist paint tool used to create halos, fog, clouds, etc. in paint programs. Most programs let you control the size and shape of the application area. Some packages provide a transparency adjustment that adjusts the density of the color.

ALAP AppleTalk Link Access Protocol. In an AppleTalk network, this link access-layer (or data link-layer) protocol governs packet transmission on LocalTalk.

Algorithm Prescribed set of mathematical steps which is used to solve a problem or conduct an operation.

Alias A feature of the Apple Macintosh System 7 allowing the user to create a file that points to the original file. When you click on an alias, the original application is launched. Aliases can work across a network; so you can access a program residing on a file server or a Mac that runs System 7 file sharing.

Aliasing Condition when graphics, either constructed with lines (vectored) or dots (bitmapped), show jagged or stair-stepped edges under magnification. Contrast with anti-aliasing.

Aligned screens Halftone screens of CYMK printed at identical angles to each other, used to reduce objectionable dot patterns that degrade image quality. See screen angles.

All Point Addressability The ability in a layout program to place text or graphics anywhere on the page. Some programs are restricted to row/column restrictions.

Alley The space between columns in a multi-column layout, or between the columns in tabular material.

Allocate To reserve the required amounts of a resource, such as disk space.

Alpha The "top" 8 bits in a 32-bit color scan or artwork. The first 24 bits describe the color of the objects; the "alpha channel" describes the object's opacity. This allows images to be layered atop one another, with some "see-through" of each one. Requires heavy processing. Present in the best imaging editing software. In video, the opacity information is carried in a component called alpha channel.

Alphabet length The width of a complete lowercase alphabet in a font and size. Used to compare fonts and plan layouts.

Alphamosaic A method of displaying very low-resolution images with elementary characters. Also known as alphageometric.

Alphanumeric Set of characters composed of letters and numbers; may include punctuation marks and other symbols; excludes printer control characters such as Carriage Return and flow control characters such as XON and XOFF.

Alphanumeric COM Computer output microform which is limited to receiving and recording letters, digits and punctuation characters. Cannot handle raster or vector graphics.

Alternate characters Different versions of a character in the same font, with stylistic changes such as extra swirlies and curlicues (technical words) to use in arty typography.

AM Amplitude Modulation. A method of adding information to an electronic signal by varying the height of its sine wave. "Modulation" means imposing information on an electrical signal. In LANs, the change in the signal is registered by the receiving device as a 1 or a 0.

Ambiguity resolution The process in OCR that uses multiple information sources to make the best choice for a character or word from several alternatives.

Ambiguous term A word that can have two meanings — like "lead" and "lead." Complicates the practice of full-text searching by concept. If you asked for documents with the word "lead" (meaning the metal) you might get hits on documents with "lead" (meaning to guide the way). Thesauruses overcome this problem.

Moore's Imaging Dictionary

Ammonia The familiar chemical compound is used in the processing of diazo film. Ammonia alkalizes (neutralizes) the acidic elements in diazo coating, leaving behind the (usually negative) image. Concern over the health implications of its use has prompted a search for an alternative.

Amp Ampere. Unit of electrical current, or rate of current flow. One volt of potential across a one ohm impedance causes a current flow of one amp. Amperage is mathematically equal to watts divided by volts.

Ampersand The "&" character. Originally was the Latin word "et," meaning "and" and usually typeset as a ligature. Has been corrupted over the years to its present form. Does not means the same as "plus."

Analog Comes from the word "analogous," meaning "similar to." Analog devices record or monitor real world happenings — motion and sound, for instance — and convert them into "analogous" electronic representations — film or audio tape, in our example. Analog means recreating the continuous nature of the original "thing"; it's the opposite of digital, which translates the original happening into ones and zeros — an "unanalogous" representation. See digital.

Analog monitor Video monitor that accepts an analog signal from the computer (digital to analog conversion is performed in the video controller). Analog monitors can be designed to accept a narrow range of display resolutions (for example, only VGA or VGA and Super VGA), or multisync analog monitors can accept a wide range of resolutions including TV (NTSC). Color analog monitors accept separate red, green and blue (RGB) signals for sharper contrast. Contrast with digital monitor.

Analog video A video signal that represents an infinite number of smooth gradations between colors and luminance levels.

Anamorphic Unequally scaled in the vertical and/or horizontal dimension.

Animated graphics Moving diagrams or cartoons. Often found in computer-based courseware, animated graphics take up far less disk space than video images.

Annotation The ability to attach notes to graphics or images by typing them in, using a light pen or digitizing tablet. Useful for clarifying documents or editing images. Microsoft's OLE in Windows introduced the idea of audio annotation, i.e., spoken or recorded comments.

ANSI American National Standards Institute. A standards-setting, non-government organization, which develops and publishes standards for "voluntary" use in the United States. Standards set by national organizations are accepted by vendors in that country. ANSI is located at 1430 Broadway, New York NY 10018 212-642-4900.

ANSI character set The American National Standard Institute 8-bit character set. It contains 256 characters.

Anti-aliasing Blending techniques that smooth the jagged edges of computer generated graphics and type. A common anti-aliasing technique is to fill the pixels between the jagged ends with levels of gray or color to soften the edge and blend it smoothly into the background.

Anti-halation backing Coating on the back surface of film that absorbs light, preventing the reflection that causes a "halo" or glowing effect that reduces resolution.

Antique finish Paper treated in such a way — with a rough surface, yellowing color and visible pulp — to simulate aging.

Any key Some programs and batch files pause, asking for your permission to continue by pressing "any" key. Technical support operators tell us many people actually call saying "I've looked all over my keyboard, and my computer doesn't have an "any" key."

A-O Acoustic-Optic. Using sound waves passing through a transparent medium to deflect a laser beam. Used in newer laser COM recorders.

APA All Points Addressable. Refers to an array (bit-mapped screen, matrix, etc.) in which all bits or cells can be individually manipulated.

Aperture An opening which lets light through. A smaller or larger aperture controls the amount of light reaching film or the collection matrix of a CCD.

Aperture card Paper card the size of an IBM punch card with a rectangular opening that holds a 35mm frame of microfilm. Retrieval information can be punched into the card.

Moore's Imaging Dictionary

Aperture fluorescent lighting A scanner designs that uses a fluorescent light source with a narrow unfrosted strip that lets brighter light pass through with each scan — which boosts the accuracy of the captured data.

Apex The point of a character where two lines meet, such as the top of an "A."

API Application Program Interface. Generic term for any language and format used by one program to help it communicate with another program. Specifically, an imaging vendor can provide an API that enables programmers to repackage or recombine parts of the vendor's imaging system, or integrate the imaging systems with other applications, or to customize the user interface to the imaging system. Some commonly known APIs are NetBIOS, Berkeley Sockets and Named Pipes.

Append To add text or data to the bottom of a file. Usually the alternative option to "overwriting" — replacing — the original file.

Apple Apple Computer, Inc., Cupertino, CA. Manufacturer of personal computers. Heavy penetration in the graphics/desktop publishing business. Apple was formed on April Fool's Day, 1976, by Steve Wozniak and Steve Jobs, aided greatly by Mike Markkula.

AppleShare Apple software that turns a Macintosh into a file server. Also, allows non-Apple computers to work on an Appletalk network.

AppleTalk Apple's LAN, it supports Apple's LocalTalk as well as Ethernet and Token Ring. Built into all Macs and LaserWriters. can support non-Apple computers.

Application A broad and generic term for any software program that carries out some useful task. Word processors and graphics programs are applications.

Application generator Software that generates an executable program from information supplied by the user. Sort of an automatic programmer. Think of the irony: Some computer programmer made himself obsolete by inventing an application generator!

Application server a server dedicated to providing specific programs to users on a network.

ARC One of the first data compression utilities. Used for archiving files. From System Enhancement Associates, Inc., Clifton, NJ. Has been upgraded to ARC+Plus.

Archie Located on several computers around the country, Archie is a kind of superdirectory to the files on the Internet. Search for files or by subject.

Architecture Refers to the way a system is designed and how the components are connected with each other. There are computer architectures, network architectures and software architectures.

Archive A copy of data on disks, discs, CD-ROM, mag tape, etc., for long-term storage and later possible access. Archived files are often compressed to save storage space. Contrast with backup. Used as a verb, it means the act of "migrating" data from an on- or near-line storage device to off-line storage.

Archival quality The extent to which a reproduced image will (or won't) last "forever." Though not a precise standard, archival quality media should hold data safely for 50 years.

Area composition Positioning text and art onto a page for reproduction. In other words, layout. In the old days, it was called paste-up.

Array An orderly arrangement of devices, components or data elements that combine to do one task. A CCD has an array of photosensors. A RAID has an array of disk drives.

Arrhenius Lifetime Plot A totally weird name for a graphical means of estimating the life of any material, by charting the end of the material's "life" at super-high temperatures, and interpolating how long it would last at a more normal temperature. Hot stuff. Used to "age" optical discs.

ART Adaptive Recognition Technology. Calera's (Sunnyvale, CA) OCR technology that uses information gathered about the text to make educated guesses to recognize words. Example: The letters "MA - - - L" (with the middle letters unrecognizable) could be either the word "manual" or "mammal." ART could note that the word "mammal" has appeared elseswhere several times in the document, so chances are good it's "mammal."

Artifact Defect or undesirable element in an image or video picture. that is caused by the system's limitation.

Artificial intelligence AI. The ability of computers to adjust their processes in response to new input. Said to resemble the learning and reasoning behavior of humans. Maybe. About as much as virtual reality resembles reality. What AI does is allow computers to adjust and restructure databases dynamically ("on the fly") as new relationships among data let the computer draw different conclusions. For example, in OCR a group of AI-based subroutines can improve their confidence in a character presented in a new way by accepting new "assumptions" about the character's shape into their memories. This is how handprinting OCR (or, ugh, "ICR") works.

Ascender A typographic term for the portion of lowercase characters that rises above the main body of the letter. The lowercase letters b, d, f, h, k, l and t have ascenders.

ASCII American Standard Code for Information Interchange. Pronounced AS-kee. It's the most popular coding method used by small computers for converting letters, numbers, punctuation and control codes into digital form. Once defined, ASCII characters can be recognized and understood by other computers and by communications devices. ASCII represents characters, numbers, punctuation marks or signals in seven on-off bits. A capital "C", for example, is 1000011, while a "3" is 0110011. Seven-bit encoding allows it to represent 128 symbols. Eight-bit ASCII encoding — so called extended ASCII — extends ASCII's symbols to 256.

This compatible encoding (it was developed by ANSI — the American National Standards Institute) allows virtually all personal computers to talk to each other, if they use a compatible modem or null modem cable, and transmit and receive at the same speed.

ASCII sort A means of alphabetizing that accounts for capital letters and numbers. To arrange something in an ASCII sort, numbers (digits) come first in numerical order, followed by capital letters in alphabetical order, followed by lower case lower case characters in alphabetical order. This glossary is NOT in an ASCII sort.

ASIC Application Specific Integrated Circuit. A chip customized with application software permanently written in (firmware) that performs a particular function.

Aspect ratio The relationship of width to height. When an image is displayed on different screens or on paper or microform, the aspect ratio must be kept the same. Otherwise the image will be "stretched" either vertically or horizontally.

For example, 9" x 12" screen has an aspect ratio of 3:4 (said three-to-four). 10" x 30" has an aspect ratio of 1:3. A TV screen is three quarters as high as it is wide, so it has a 3:4 aspect ratio.

ASPI Advanced SCSI Programming Interface, which acts as a liaison between SCSI device drivers and the interface card (also known as the host adapter).

Whenever a new device is added to a computer system, a software program called a "driver" must tell the computer how to talk to the new device. Instead of forcing vendors to write drivers for every host adapter, ASPI lets them write a driver to ASPI standards, supposedly guaranteeing that the device the driver controls will work with all ASPI-compatible host adapters.

Assemble edit The simultaneous recording of the video, audio and control tracks on a segment of videotape. Assemble edits are usually performed at the VTR.

Associate A Windows NT term. To identify a filename extension as "belonging" to a certain application, so that when you open any file with that extension, the application starts automatically.

Asymmetrical compression. Any compression technique that requires a lot of processing on the compression end, but little processing to decompress the image. Used in CD-ROM creation, where time and costs can be incurred on the production end, but playback must be inexpensive and easy.

Asynchronous Literally, not synchronous. A method of data transmission in which characters are sent at irregular intervals by preceding each character with a start bit, and following it with a stop bit. Used by most small computers (especially PCs) to communicate with each other.

ATL Automated Tape Library. Large-scale tape storage system, which uses multiple tape drives and mechanism to address 50 or more cassettes.

ATSC Advanced Television Systems Committee. An organization created to coordinate the standards applied to the transmission, generation and reception of high definition television.

Atto A numerical prefix describing a power of 10^{-18}.

Attribute In graphics, the condition a font is in — i.e., boldface,

Moore's Imaging Dictionary

italic, underlined, reverse video — is its attribute. In MS-DOS, files can be assigned attributes that define how accessible it is — i.e., "read-only" is a file's attribute. Attributes include: read-only, hidden, system file, changed since backup. In a document retrieval system, an attribute of a file is one of the keys by which the document has been stored and indexed.

Audio What a human can hear. Audio frequencies range from 15Hz to 20,000Hz.

Audio mix The audio signal resulting from the modification or combination of one or more audio sources.

Audio state An audio setting which determines the volume level for one or more audio sources. Crossfade transitions are performed between one audio state and another.

Audiovisual Output you can see and hear. Television is audiovisual. Radio is not.

Audit trail Record of activity that has occurred in a certain file, or on a certain computer.

Authoring Software that helps multimedia developers create interactive presentations without much programming overhead. Allows you to incorporate many data types and also gives you control needed to play back information on CD-ROMS or hard disks or whatever.

Authorization code Identifying code — often a password — that allows a user access to a system. Used mainly for privacy and security. Also used to divide up a computer's capacity among departments and/or hierarchies; different "grades of service" are given different authorization codes.

Auto assembly Recording a final edited videotape using an Edit Decision List (EDL) generated by a computerized video editing system.

AutoCAD One of the first computer-aided-design programs, by AutoDesk Inc., Sausalito, CA. It works on PCs, VAXs, Macs and Unix workstations.

Autochanger A device that holds multiple optical discs and one or more disc drive. Associated software can swap discs in and out of the drive as needed. AKA jukebox.

AutoLISP AutoCAD's language for create customized menus and routines.

Automated retrieval Using a computer to identify and locate a stored image of some kind. Generally requires the use of key words or codes in an indexing scheme.

Automatic blue frame In painting, animation and imaging editing programs, the ability to display a faint blue copy of the previous version or cel while editing the current image to match positions precisely or create an accurate trace.

Automatic decimal tab. The ability of a wordprocessor to align numerical values vertically along the decimal points. Saves the typist from manually aligning columns of dollar figures.

Automatic Document Feeder. See ADF.

Automatic file select The ability of software to retrieve certain records from data files based on the appearance of certain characters in a certain field. This can pull records with certain zip codes in the address field.

Autoreversing/Autopositive Types of COM film which yield a negative image (white characters on black background). Users prefer negative images because they're easier to see in a microfiche reader.

Auto selection The ability of photo manipulation/imaging editing software to select entire areas of the image within a specified range of color values.

Autostitch The ability of a hand-scanner's software to combine several passes over a document into a single image.

Auxiliary storage External storage devices, such as disk drives, optical drives and tape recorders. Can hold more than the main memory (i.e. RAM), but has far slower access times.

Average Latency The time required for a disk or disc to rotate one-half revolution. The idea is that when a drive finds the track it wants, it has to wait an average of one-half rotation before the sector it wants comes around.

AVC Automatic Volume Control.

AVL Automated Vehicle Location. With satellites and GIS (Geographic Information Services), it's possible to plot the location and movement of cars and trucks.

B The second letter of the alphabet. In ASCII, uppercase "B" is represented as hexadecimal 42; a lowercase "b" is hexadecimal 62. In EBCDIC, an uppercase "B" is hexadecimal C2; a lowercase "b" is hexadecimal 82.

B In video, one of the three primary color signals (red, green and blue) that are produced by a video camera or applied to a monitor.

Backbone The backbone is the part of the communications network which carries the heaviest traffic. The backbone is also that part of a network which joins LANs together — either inside a building or across a city or the country. LANs are connected to the backbone via bridges and/or routers and the backbone serves as a communications highway for LAN-to-LAN traffic.

Backfile conversion The process of scanning in, indexing and storing a large backlog of paper or microform documents in preparation of an imaging system. Because of the time-consuming and specialized nature of the task, it is generally performed by a service bureau — called "outsourcing."

Background (1) The simultaneous, non-interrupting, execution of an automatic program while the computer is being used for something else.

(2) The portion of microfilm that doesn't have anything recorded on it. It may be opaque or clear, depending on whether the film has a nega-

tive (background is opaque) or a positive (background is clear) image.

(3) The area of your Windows desktop behind and around your windows and icons. The color and pattern you put on it through the desktop manager is called wallpaper.

Background ink A reflective ink used to print the parts of a document that are not meant to be picked up by a scanner or optical character reader.

Background program A low-priority program operating automatically when a higher priority (foreground) program is not using the computer's resources.

Background task A secondary job performed while the user is performing a primary task. For example, many network servers will carry out the duties of the network (like controlling who is talking to whom) in the background, while at the same time the user is running his own foreground application (like word processing).

Backlit Any screen that has a light source which shines from the back of the image toward the viewer, making image sharper and easier to see in low ambient lighting conditions.

Backoff When a device attempts to transmit data and it finds trouble, the sending device must try again. It may not try again immediately. It may "back off" for a little time so the trouble on the line can be cleared. This happens with LANs. For example, an earlier attempt to transmit may have resulted in a collision in a CSMA/CD (Carrier Sense Multiple Access/Collision Detection) Local Area Network (LAN). So the device "backs off," waits a little and then tries again. How long it waits is determined by preset protocols.

Backplane A printed circuit card located on the back of a rack or a computer chassis, which has sockets into which expansion boards are connected.

Backslash Also called a virgule, the backslash key achieved fame because Microsoft used it to bring distinguish between subdirectories in MS-DOS.

Backup As a noun, it's a duplicate copy of data placed in a separate, safe "place" — electronic storage, on a tape, on a disk, in a vault — to guard against total loss when — not "if" — original data somehow becomes inaccessible. Generally for short-term safety.

Contrast with "archive," which is a filed-away record of data meant to be maintained a long time, in the event of future reference. As a verb, to "back up" means to physically make the copy. Also, in printing it means to print on the back of a sheet that's been printed on one side.

Back up server A program or device that copies files so at least two up-to-date copies always exist.

Bad block A defective area on a storage medium that software cannot read or write.

Bad break In typography, when a word is incorrectly hyphenated and the second half moves to the next line. Also when one word or part of a hyphenated word falls onto a line by itself (called a "widow").

Bad sector Defective areas on a floppy or hard disk. The MS-DOS "Format" command recognizes bad sectors, and "locks them out" so the computer won't try to place any data on those sectors.

Baffle A partition inside a loudspeaker to prevent air vibrations striking the back of the speaker's diaphragm from cancelling out the vibrations from the front of the diaphragm. Particularly valuable in the reproduction of bass notes.

Ballistic gain Feature on some trackballs and mice that makes the cursor move in relation to the speed the user moves the mouse or trackball. The faster the mouse, the farther the cursor moves. Move the mouse slowly, and the cursor moves only a tiny bit.

Balloon help The on-screen help instructions associated with the Macintosh System 7. The dialog box is displayed in a talk balloon, like in a cartoon.

Balun Contraction of BALanced/UNbalanced. An impedance matching transformer. Baluns are small, passive devices that convert the impedance of coaxial cable so that its signal can run on twisted-pair wiring. They are often used so that IBM 3270-type terminals, which traditionally require coaxial cable connection to their host computer, can run off twisted-pair. Works for some types of protocols and not for others. There is often some performance degradation with baluns. And the signal cannot run as far on twisted wire as it can on coaxial cable. But twisted pair is cheaper.

Moore's Imaging Dictionary

Bakelite An obsolete insulating material of the phenolic (synthetic resin) group. Jewelry made of bakelite is now particularly prized.

Band 1. Strip of metal or mylar that has characters on it, which is used by a type of printer (called a band or belt printer) to print on impact. 2. 32 scans lines of video data.

Band buffer Temporary storage where a band of video output (32 scan lines) is held before being fed to a laser-based printer.

Band, frequency The frequencies between the upper and lower bands. These are the accepted band frequencies and their acronyms:

Below 300 Hertz	ELF	Extremely low frequency
300—3,000 Hertz	ILF	Infra Low Frequency
3—30 kHz	VLF	Very Low Frequency
30—300 kHz	LF	Low Frequency
300—3,000 kHz	MF	Medium Frequency
3—30 MHz	HF	High Frequency
30—300 MHz	VHF	Very High Frequency
300—3,000 MHz	UHF	Ultra High Frequency
3—30GHz	SHF	Super High Frequency
30—300GHz	EHF	Extremely High Frequency
300—3,000 GHz	THF	Tremendously High Frequency

Bandwidth The range of frequencies, express in Hz, that can pass through a circuit.

Bandwidth on demand Just what it sounds like. You want two 56 Kbps circuits this moment for a videoconference. No problem. Use one of the newer pieces of telecommunications equipment and "dial up" the bandwidth you need. Uses for bandwidth on demand include videoconferencing, LAN interconnection and disaster recovery. Bandwidth on demand is typically only for digital circuits and it's typically carved out via a T-1 permanently connected from a customer's premises to a long distance carrier's central office, also called a POP — Point of Presence.

Bar code A system of portraying data in a series of machine-readable lines of varying widths. The "UPC" on consumer items is a bar code. In document management, a bar code is used to encode indexing information.

In microfiche, bar codes allow the automatic control of the duplication process, plus contain indexing information. These bar codes usually appear in the last two or three title frames in the first title row of a microfiche.

Barium ferrite A type of magnetic particle used in some recording media including floptical diskettes.

Barrel distortion When the horizontal and vertical lines on a monitor are bowed out at the center.

Base font Typeface that graphics software defaults to if no other font is specified.

Base alignment Arrangement that allows columns of text to fall on the same line across the page, regardless of varying sizes of the elements in the columns.

Baseline The imaginary horizontal line upon which typeset characters appear to rest.

Batch file A text file for personal computers (with the extension .BAT) that contains MS-DOS commands. When you launch the file (by typing its name without the extension), DOS carries out the commands in the file. Batch files simplify life. You can save yourself from typing things over and over. You also don't have to remember all the commands to do something, just the first one.

Batch processing Conducting a group of similar computer tasks at one time, instead of steadily throughout the day.

Batch program An ASCII file (text title) that contains one or more commands. When you run a batch program, the commands are processed sequentially. In DOS and Windows, a batch program's filename has a .BAT extension.

Bates coding A filing scheme used by lawyers to organize documents according to their strategy. Problem: makes it difficult to change strategies mid-stream. Imaging-based document management solves the problem, and gets the lawyers onto a new strategy quicker. Saves money, wins cases.

Moore's Imaging Dictionary

Baud A measure of transmission speed over an analog phone line — i.e. a common POTS line. (POTS stands for Plain Old Telephone Service). Imagine that you want to send digital information (say from your computer) over a POTS phone line. You buy a modem. A modem is a device for converting digital on-off signals which your computer speaks to the analog, sine-wave signals your phone line speaks. For your modem to put data on your phone line means it must send out an analog sine wave (called the carrier signal) and change that carrier signal in line with the data it's sending. Baud rate measures the number of number of changes per second in that analog sine wave signal. According to Bell Labs, the most changes you can get out of a 3 KHz (3000 cycles per second) voice channel (which is what all voice channels are) is theoretically twice the bandwidth, or 6,000 baud. Baud rate is often confused with bits per second, which is a transfer rate measuring exactly how many bits of computer data per second can be sent over a telephone line. You can get more data per second — i.e. more bits per second — on a voice channel than you can change the signal. You do this through the magic of coding techniques, such as phase-shift keying.

Baudot Code The code set used in telex transmission, named for French telegrapher Emile Baudot (1845-1903) who invented it. Also known by the CCITT approved name, International Telegraph Alphabet 2. Baudot code has only five bits, meaning that only 32 separate and distinct characters are possible from this code, i.e. 2 x 2 x 2 x 2 x 2 equals 32. By having one character called Letters (usually marked LTRS on the keyboard) which means "all the characters that follow are alphabetic characters," and having one other key called Figures (marked FIGS), meaning "all characters that follow are numerals or punctuation characters," the Baudot character set can represent 52 (26 x 2) printing characters. The characters "space," "carriage return," "line feed" and "blank" mean the same in either FIGS or LTRS.

BBS Bulletin Board System. Another term for an electronic bulletin board. Typically a PC, modem/s and communications bulletin board software attached to the end of one or more phone lines. Callers can call the BBS, read messages and download public domain software. IMAGING MAGAZINE has a BBS called the InfoBoard. It's open to our readers and friends. Its number is 212-989-4675. Settings to call this number are 8 bit, no parity, one stop bit, 2400 baud. The person who operates a BBS is called a SYSOP (pronounced "sis-op").

Moore's Imaging Dictionary

BCC Block Check Character. In data transmission, a control character appended to blocks in character-oriented protocols and used for figuring if the block was received in error. See CRC.

Beam recording Using an electron or laser beam to record directly onto film.

Beam splitter Divides a light beam into two or more separate beams. Used in color separation and film recorders.

Benchmark A standardized task to test the capabilities of various devices against each other for such measures as speed.

BER Bit error rate. A measurement of the average number of errors which occur (or can occur) while writing or transmitting data.

Bernoulli Box A storage disk system that uses the principles of fluid dynamics (discovered by 18th century Swiss scientist Daniel Bernoulli). When the disk rotates at great speed, it creates a cushion of air, keeping the read/write head at the perfect close distance from the disk's surface. Capacity depends on the cartridge capacity, which range from 10 to 90 megabytes.

Beta 1. 1/2" analog component videotape recording format developed by Sony which records luminance at full bandwidth and color difference signals at half bandwidth. Superceded in the consumer market by VHS, "Beta" is widely used in commercial video production.

2. A beta is the final version of a new product before its release.

Bezel The metal or plastic frame that surrounds a display tube.

Bezier curve Mathematically defined curve, used in CAD and graphics application software to create curved images. A Bezier curve is made up of only four points: the two ends and two other points that affect its shape. Contrast with spline.

BFT Binary File Transmission. Standard for transmitting facsimile data between fax boards directly. Faster than conventional fax modems.

Bidirectional printing A typewriter always prints from left to right. So did the early computer printers. That's unidirectional printing. The newer computer printers will print from left to right, drop down a line, then print from right to left. Bi-directional. This increases the printer's speed.

Moore's Imaging Dictionary

Bidirectional symbol A bar code that can be read either left to right or right to left.

Bifurcate To divide in two.

Bilevel A binary scan that assigns each pixel an attribute of either black or white — no gray tones, no colors.

Billibit Someone's absolutely awful term for one billion bits. Also (and better) called a gigabit.

Binary Where only two values or states are possible for a particular condition, such as "ON" or "OFF" or "One" or "Zero." Binary is the way digital computers function because they can only represent things as "ON" or "OFF." This binary system contrasts with the "normal" way we write numbers — i.e. decimal. In decimal, every time you push the number one position to the left, it means you increase it by 10. For example, 100 is 10 times the number 10. Computers don't work this way. They work with binary notation. Every time you push the number one position to the right it means you double it. In binary, only two digits are used — the "0" (zero) and the "1" (the one). If you write the number 10101 in binary, and you want to figure it in decimal as we know it, here's how you do it. 1 is one thing; Zero x 2 = zero; 1 times 2 x 2 = 4; 0 x 2 x 2 x 2 = 0; 1 x 2 x 2 x 2 x 2 = 16. Therefore the total 10101 in binary = 1 + 0 + 4 + 0 + 16 = 21 in decimal.

Binary digit Long for bit.

Bindery The information about a Novell NetWare users' access rights. A database containing definitions for entities such as users, groups and workgroups. The bindery contains three components: objects, properties, and property data sets. Objects represent any physical or logical entity, including users, user groups, file servers, print servers, or any other entity given a name. Properties are the characteristics of each bindery object, including passwords, account restrictions, account balances, internetwork addresses, list of authorized clients, workgroups, and group members. Property data sets are values assigned to entities bindery properties.

BIOS Basic input/output system, which is software (usually contained in ROM) that handles the transfer of information between system elements such as memory, disks and the monitor.

bis Twice, or second version. CCITT uses the term to designate the second in a family of related standards. "ter" designates the third in

Moore's Imaging Dictionary

a family.

Bisynch Binary synchronous communications. 1. In data transmission the synchronization of the transmitted characters by timing signals. The timing elements at the sending and receiving terminal define where one character ends and another begins. There are no start or stop elements in this form of transmission. 2. Also a uniform discipline or protocol for synchronized transmission of binary coded data using a set of control characters and control character sequences.

Bit Contraction for Binary digiT. The smallest unit of data a computer can process. Represents one of two conditions: on or off; 1 or 0, mark or space; something or nothing. Bits are arranged into groups of eight called bytes. A byte is the equivalent of one character.

Bit depth Number of colors or levels of gray that can be displayed at one time. Controlled by the amount of memory in the computer's graphics controller card. An 8-bit controller can display 256 colors or levels of gray. A 16-bit can show 64,000 colors. A 24-bit controller can display 16.8 million colors or gray levels.

Bitmap Representation of characters or graphics by individual pixels, or points of light, dark or color, arranged in row (horizontal) and column(vertical) order. Each pixel is represented by either one bit (simple black & white) or up to 32 bits (fancy high definition color).

Bitmap display In a bitmap display, a screen image is generated through a 1:1 correspondence between bits in memory and pixels on the screen.

Bitmapped font A set of dot patterns that represent all the letters, characters and digits in a type font at a particular size.

Bitmapped graphics Images which are created with sets of pixels, or dots. Also called raster graphics. Contrast with vector graphics.

Bitblt Bit BLock Transfer. A feature that moves a collection of bits (usually in a rectangular shape) from memory to the display very quickly. Used to display moving objects.

Bit rate Speed at which data is transferred, expressed in bits per second. See baud rate.

Blackletter Any squared-off, sans serif typeface. AKA gothic.

Moore's Imaging Dictionary

Black line A positive image — black on a clear or white background. Opposite of "white line," also known as a negative image.

Black matrix Picture tube in which the color phosphors are surrounded by black for increased clarity.

Blend A graphics software tool. The smooth transition from one color to another. Blending tools give a realistic look to a drawing, especially if you want to create a smooth shadow.

Blind text The ability for a wordprocessor to permit a section of text to print on one hard copy, and not another.

Blip A mark placed on a microfilm for counting or timing purposes.

BLOB Binary large object. A data type that can contain any amount of binary data, and allows it to be handled as a database field. Therefore, scanned images, sound clips, video can be directly managed with a database.

Block The amount of data recorded contiguously on magnetic tape or disk in a single operation. Blocks are separated by physical gaps, or identified by their track/sector addresses.

Block move The ability for a wordprocessor to select a certain amount of text, and place it a temporary buffer in order to paste it somewhere else in the document, or delete it, or create a new file with it.

Blocking or Blockiness The ability to discern blocks or areas of pixels. An aritifact (undesirable effect) caused by JPEG.

Blooming A monitor distortion in which images enlarge as the brightness increases.

Blue collar computer A colloquial term for a handheld computer which is used by "blue collar" workers for tasks such taking inventory, tracking goods, etc. Such computer may have a pen, a large pen-sensitive or touch-sensitive screens, a bar code scanner and a modem. It may be able to capture signatures — useful for confirmation of the delivery of goods.

Blue line A printer's proof made from direct contact of negatives with a photosensitive paper. Also known as brownlines or Dylux proofs.

Blue screen Filming or videotaping an object or performer against a blue-colored background. The blue is then removed and/or replaced with either an analog or digital image, called a chroma key. The weatherman is shot against a blue screen, which is electronically replaced with a chroma key image of the weather map.

BNC A popular "bayonet-type" connector used for coaxial cables.

Board A circuit card.

Boilerplate Common material that is used over and over from one document to the next.

Boolean logic A text searching (also used in databases of all kinds) tool that identifies documents by the relationships of words within them. Uses logical "operators" like AND, NOT, OR. For example, a Boolean search could find only those documents that contain "Clinton" AND "Perot" but NOT "presidential race." Powerful way to narrow searches from a large document pool.

Boot To start your computer and have all your customized configurations initiated.

Boot partition In Windows NT, a volume that has the Windows NT operating systems and its support files. The boot partition can be (but does not have to be) the same as the system partition. It also cannot be part of a stripe set or volume set.

Bond A common, pretty-good grade of paper used in stationery and in copiers.

Bounding box A box drawn on-screen that represents the size of an object.

Boutique Slangy term for any output that demands high-quality graphics, color, stock, etc. "That annual report is a boutique job."

Bpi Bits Per Inch. Measurement of the number of bits stored per linear inch on magnetic tape. Measures density.

BPP Bits per pixels. Same as bit depth.

Bps Bits Per Second. The number of bits transferred in a data communications system. Measures speed. Don't assume that just because a LAN or other data communications system has a faster bits per second rate, it will transmit your information faster. You have to factor in speed of writing and reading from the disk and the accuracy of transmission. All datacom schemes have error-checking systems, some better than others. Typically such systems force a retransmission of data if a mistake is detected. You might have a fast, but "dirty" (i.e. lots of errors) transmission medium, which may need lots of re-transmissions. Thus, the effective bps of a "fast" data communications network may actually be quite low.

Break action Pre-set instructions that tell a COM system where to place the next block of data after a break condition.

Break condition Also called fiche break. Separating recorded frames of data on microfiche. Instructs the COM system to insert blank pages, go to the next column, go to the next microfiche, whichever, in preparation of recording the next frame of data.

Bridge A device that connects LANs using similar or dissimilar media and signaling systems such as Ethernet, Token-Ring and X.25.

Brightness The reflectiveness of an overall image, including both light and dark areas. Contrast with contrast.

Brittle Describes what happens when a rule-based OCR program is fed too many "special circumstance" rules, and the recognition engine is no longer able to recognize characters it was previously able to handle.

Broadband A communications channel with a bandwidth from 10,000 to over 1 million bits per second.

Broadcast To send a single message or fax simultaneously to many receivers.

Broadcast list A list of two or more system users to whom messages are sent simultaneously. Master Broadcast Lists are shared by all system users and are set up by the System Administrator. Personal Lists are set up by individual subscribers.

Bromide A photographic paper used in phototypesetting.

Brontobyte A really lot of memory. Bigger than megabyte, huger than a gigabyte. What you say when you have no idea of how large something is.

Browser 1. A "roadmap" to a hypertext system, usually using graphic icons, to show the available hypertext nodes and their links and relationships.

2. A software utility that displays graphic images as smaller "thumbnails" for easier retrieval.

Brush A paint package's most basic image-creation tool. Most packages let you select a variety of sizes and shapes. Many let you customize shapes.

Bruning A quality proof of typeset material produced on a Bruning photocopier.

BS Not what you think. Stands for Business Speed. A measure of quality published by printer manufacturers to describe the time from pressing the enter key to when a page is printed. Come to think of it, maybe it IS what you think.

BSC Binary Synchronous Communications. An IBM term for a communications protocol.

BTAM Basic Telecommunications Access Method. An access method that allows reads and writes between remote devices.

Btrieve A key-indexed database record management system. You can retrieve, insert, update or delete records by key value, using sequential or random access methods. First introduced in 1983, Btrieve was one of the first databases designed for LANs. Novell bought the company in the late 1980s. Now called NetWare Btrieve, it is included with most versions of NetWare 2.x and NetWare 3.x. There are Btrieve developer products for DOS, OS/2 and Windows.

BTS Burster/Trimmer/Stacker. A post-processing device at the end of certain IBM large printers.

Bubble One technique for recording data on optical discs. A laser strikes the non-image areas of the recording medium, causing bubbles to form and leaving the image area clear to reflect the reading laser's light.

Bubble jet A printer technology which works like this: A miniature electric heater is imbedded above each of the print head's group of nozzles. Electric current flows to the heater element to create a bubble. The bubble creates a disturbance in the ink chamber and forms a droplet. Thousands of droplets form an image.

Bubble memory A nonvolatile storage technique that uses pulse of magnetism to create isolated "islands" or bubbles of data.

Buffer Device or allocated memory space used for temporary storage to compensate for the difference in speed between CPUs and devices. Like a dam holds back water and trickles it out in a manageable stream. Printers commonly use buffers, for example, to hold incoming text because the text arrives at a much faster rate than the printer can output.

Built up fractions Typeset fractions where the numerator and denominator are set at the full type size, which makes it taller than the rest of the type, and requires more leading space between lines.

Bullet A round dot — a "blob" — to emphasize a brief point in printed text.

Bundling A marketing strategy where several items — e.g., software plus its accompanying hardware device — are sold together for a discounted price.

Burst 1. Sending a large amount of data, using the entire bandwidth of the channel, in one contiguous "chunk." 2. To separate continuous forms along the perforations into single sheets.

Bus Signal path or line shared by many circuits or devices. Information is often sent to all devices throughout the same bus; only the device to which it is addressed will accept it. This make designing system architecture much easier; devices can be plugged in "anywhere on the bus."

Bus mouse Mouse that takes up an expansion slot in a PC, rather that a serial port (those are called "serial mice"). There are generally more expansion slots available than serial ports.

Business graphics Transforming numerical data into charts and graphs for business analysis.

BVB A video preview technique that shows black, then the video to be edited, then black again.

Bisynch Data transmission in which synchronization is managed by timing signals generated by both ends.

Byte Common unit of computer storage. A byte is eight bits of information, one of which may be a parity bit. Generally, eight bits equals one character. Also called an "octet."

C The third letter of the alphabet. In ASCII, uppercase "C" is represented as hexadecimal 43; a lowercase "c" is hexadecimal 63. In EBCDIC, an uppercase "C" is hexadecimal C3; a lowercase "c" is hexadecimal 83.

CABSC (Canadian Advanced Broadcast Systems Committee) An organization developed by the Canadian government, Canadian Broadcast Corporation (CBS) and others to develop standards for high definition television in Canada.

Cache Pronounced "cash." Small portion of high-speed memory used for temporary storage of frequently used data. Reduces the time it would take to access that data, since it no longer has to be retrieved from the disk. You can increase the throughput of an otherwise slow-ish optical drive or jukebox with "smart caching." Software anticipates what your next request might be, and puts that into cache memory. For example, if you request page one of a document, the software assumes you may want page two next, and retrieves that also.

Cache hit When the data you want is actually in cache. Thus you don't have to access your hard disk and your computing is fast.

Cache miss When the caching software guesses wrongly and you have to read your data off your hard disk rather than reading it from the cache.

Moore's Imaging Dictionary

Caching An "intelligent" way to speed up jukebox retrievals. It is based on three principles: (1) data is retrieved in fairly predictable ways; (2) no data access is truly random; and (3) the probability of re-access is higher than the probability of initial access. Caching makes an educated guess and retrieves documents you haven't asked for yet. These are staged in your local PC or in a network server, so when you DO ask for them, they are available instantly.

CAD Computer Aided (or Assisted) Design or Drafting. It means using computers to design products or architecture. Sometimes CAD output is used to operate the machinery that makes the product; this is called Computer Aided Manufacturing. See CAM. CAD design is based on vector graphics which can be mathematically manipulated for precision. Also stands for Computer Aided Dispatching.

CADAM IBM CAD software.

Caddy The shell of an optical disc. Protects it from grubby fingerprints, and includes write protection devices. AKA case.

CADE Computer Aided Data Entry. Using the check marks and hand printing on scanned or faxed forms to automatically fill in databases. For certain well-planned applications (and when the form is designed right), it eliminates manual key entry altogether. I first heard the acronym from DataCap, who makes the fabulous Paper Keyboard CADE product.

CAEDS IBM CAD software.

Call In programming, to invoke a subroutine or function.

Callback modem A modem that calls you back. Here's how it works. You dial into a network. A modem answers. You put your password in. It accepts the password. It says "Please hang up. I will now call you back." You hang up. It calls you back. There are two reasons for doing this instead of allowing you to just go straight into the network. 1. It's better security. You have to be at a predetermined place — an authorized phone number. 2. It may save on phone calls. The modem uses the company's communications network, which is probably cheaper than what the person calling in can use.

Callout Text pulled from within the body and typeset in a graphic way to illustrate a point. Also called a pull-quote.

CALS Computer-aided Acquisition & Logistics Support. The Defense Department's initiative to make vendors comply to standards.

CAM Computer Aided (or Assisted) Manufacturing. This is when machinery (often robotics) to build products gets its instructions from computer input. This input usually originates in a CAD device. The CAD output — the design for a new widget — is fed into the CAM device and translated into instructions to the machinery that makes the widget. When they are integrated this way, it is known as CAD/CAM.

Camera ready The final pre-processing step in graphics, resulting in hard copy or negatives to be used by the printer.

CANDA Computer Assisted New Drug Application. The FDA requires a ton of documentation for an NDA — New Drug Application. When you automate the document management, imaging and retrieval with computer technologies — such as OCR, full-text searching, boolean operators — it is a CANDA.

Cap Capital (uppercase) letters. "This headline should be caps and lowercase."

Caps and small caps Typography in which capital letters are set their normal way, and lowercase letters are set in uppercase, but smaller. It looks like this: CAPS AND SMALL CAPS.

Capacitive Touchscreen that uses a glass panel with a clear conductive coating. Works by sensing the slight electrical impulses from your finger.

Capstan The cylinder which moves a magnetic tape at the proper speed. Also used in imagesetters to pull paper or film through, exposing the image one line at a time. Not as precise as drum-based imagesetters.

CAR Computer Assisted Retrieval. Computer systems that locate or identify data stored on microform or paper. CAR systems rely on indexing and cross-indexing, pre-assigned to the documents, to find all documents related to the CAR search "attributes." A (simplified) attribute string may be something like "Purchase orders for Widget Style 007 in October." The CAR software would search "purchase orders," "Widget Style" records and "October" activity files for matches. It would then notify the user of the whereabouts of the relevant documents.

The arcana of document management system design — for instance: whether documents are kept redundantly under all possible categories; whether the document's index simply identifies the location of the document or actually summarizes its contents — is beyond the reach of this glossary. Best advice: Subscribe to IMAGING MAGAZINE.

(2) Courtesy Amount Reading. Scanner/OCR software systems that read the handwritten numeric dollar amounts from checks.

Card Printed circuit card. They carry special circuits for specific functions. They fit into PCs' expansion slots.

Carriage In jukeboxes, the "hand" on the robotic "arm" that grasps and moves the disc to or from the storage slot, disc drive or mailslot. Also called the "picker," which is more fun. Also the mechanism in a printer that moves the paper past the print head.

Carriage paper width The limit of the width of a line in word-processing. The carriage paper width is usually two inches less than the paper width — one inch on each side.

Carriage return The act of moving from one line to the next. Accomplished either automatically by word wrapping, or manually by pressing the carriage return key, usually labeled "enter."

Cartridge Generally, a removable unit which makes the replacement of some resource convenient for the user. Technically, a bullet is a cartridge — it combines the shell, gunpowder, padding, etc., into a unit that is easier to replace than each of its components would be.

For our purposes, any removable storage device that is not a cassette. In optical discs, it's called the "platter."

Cartridge font A font stored in a cartridge that physically plugs into your printer. Font cartridges are often used by HP LaserJet printers, but some dot-matrix printers also use font cartridges.

CAS Communicating Applications Specification. A high-level API (application programming interface) developed by Intel and DCA, introduced in 1988. CAS enables software developers to integrate fax capability and other communication functions into their applications.

Cascade When windows cascade, they are arranged in an overlapping pattern so the title bar of each window remains visible.

CASE Computer Aided Software Engineering. CASE is a new, faster, more efficient way of writing software for some applications. The idea with CASE is to sketch out relations between databases, events and options and then have the computer write the code.

Case 1. One "step" or path in a workflow procedure. 2. Whether a letter is a capital (uppercase) or lowercase letter. Comes from the old days when type was stored in wooden cases, with capital letters stored in the top — upper — case above the small letters, stored in the lower case.

Case fractions Fractions typeset so the numerator and the denominator are about half their normal size, allowing the fraction to fit in one line space. Also called "piece fractions."

Case sensitive Knows the difference between capital letters and lower case letters. A case-sensitive search for "CASE" would not find "case".

Cassette A magnetic tape storage device, which combines the tape, and supply and take-up reels into a removable unit. The 1/4" audio cassette and the VHS and Beta videotape units are the most familiar forms of cassette. See cartridge.

Casters to controllers Means the manufacturer made the entire machine, from top to bottom, instead of OEMing parts from other makers.

Casting off Estimating how much text will be necessary to fit a space.

CAT Computerized Axial Tomography. CAT scans are like X-rays of the brain. They are thin cross-section slices, whose images are later reassembled for diagnosis.

Catalog Another name for a listing of directories or files stored on a computer or disk.

Cathode ray tube See CRT.

CATIA IBM CAD software.

CATV Community Antenna Television. Data transmission based on radio frequencies (RF) and delivered over 75 ohm coaxial cable. Has frequency-divided channels, which makes it great for simultaneously sending data and video, and for delivering many channels of broadcast TV to your home.

CAV 1. Constant Angular Velocity. On optical discs, data is written on tracks which are concentric circles like on a magnetic disk. Used for random access storage and interactive applications. For interactive video, one frame equals one rotation at a constant 1,800 RPM rotation speed. See CLV.

2. Component Analog Video. A video format in which three separate video signals represent luminance and color information. Each signal consists of an analog voltage that varies with picture content. Also called analog component.

CBC Canadian Broadcast Corporation.

CBEMA Computer and Business Equipment Manufacturers Association.

CBT Computer-Based Training. Using the computer to teach. Also called "courseware."

CCD Charge Coupled Device. The "eyes" of a scanner or "digital camera." CCDs are small electronic devices with arrays of light-sensitive elements. The number of these elements and the width determine the scanner's resolution. Light is bounced off the image onto the CCD, which translates the varying intensities of the reflected light into digital data. CCD technology is used also in "digital still cameras" such as the Sony Mavica. The small size of the array itself — approximately microchip size — and the high resolution — around 1,000 by 1,018 elements — of these cameras have greatly improved "image acquisition" capabilities and opened up exciting new applications in manufacturing quality control and in medicine.

CCIR Comite Consultatif International des Radio-Communications. An organization formed by the United Nations to regulate international communications.

CCITT Comite Consultatif Internationale de Telegraphique et Telephonique. In English, it means the International Telephone and Telegraph Consultative Committee. It issues recommendations adopted as standards by most people and most firms.

CCS Continuous composite servo. A provision in ISO 10089 optical standard accepts track-head alignment by means of optically matching spiral grooves in the disc surface. CCS is that means.

CD Compact Disc. A standard medium for storage of digital audio data, accessible with a laser-based reader. CDs are 12 centimeters (about 4 3/4") in diameter. CDs are faster and more accurate than magnetic tape for audio. Faster, because even though data is generally written on a CD contiguously within each track, the tracks themselves are directly accessible. This means the tracks can be accessed and played back in any order. More accurate, because data is recorded directly into binary code; mag tape requires data to be translated into analog form. Also, extraneous noise (tape hiss) associated with mag tape is absent from CDs.

CD-I Compact Disc Interactive. A compact disc format, developed by Philips and Sony, which provides audio, digital data, still graphics and limited motion video.

CD-R Recordable CD disc.

CD-RDx Compact Disc Read-only Data eXchange. A proposed standard for full-text retrieval from CD-ROMs.

CD-ROM Compact Disc Read Only Memory. A data storage system using CDs as the medium. CD-ROMs usually hold 680 megabytes of data. User capacity (total minus overhead) is 556 megs. Used for publishing and distributing data. Has the same CLV (constant linear velocity) spiral format as audio CDs.

CD-ROM XA Compact Disc Read Only Memory eXtended Architecture. Microsoft's extensions to CD-ROM that let you interleave ADPCM-digitized audio with data. CD-ROM XA reads discs in all the other CD formats, making it the closest thing to a CD-ROM standard there is.

CDTV Commodore Dynamic Total Vision. Consumer multimedia system which includes a CD-ROM/CD player, 68000 processor, 1 meg of RAM and a remote control.

CD-V A compact disc format that combines full-motion video with digital audio.

Cel Short for celluloid. In traditional animation, clear sheets of plastic (cels) were stacked on top of one another to form a complete picture (or frame) and the photographed. In computer-based animation programs, a cel is an image clipped from all or part of a frame. Some animation programs can create multiframe cels.

Cell The area in a spreadsheet into which you enter data. Cells are arranged in columns and rows.

Centralized Processing All or substantially all of an enterprise's computing is done in one site, usually called the data center. Was the norm in US business until the penetration of desktop PCs and local area networks, which led the way to distributed processing.

Centronics De facto standard 8-bit, 36-pin parallel interface for connecting printers and other devices to a computer.

CEPS Color Electronic Prepress System. A computer-controlled imaging and pagination system for color printing. If the color computer is digital, it uses the Neugebauer equations to calculate color values.

CFS Contextual File Search.

CG Character Generation. In video, the electronic creation of letters, numbers, words and other characters to be superimposed on a video stream.

CGA Color/Graphics Adapter. A low resolution video display standard, invented for the first IBM PC and now essentially obsolete.

CGI Computer Graphics Interface. Graphics standard for display screens, printers and plotters. Superceded GKS.

CGM Computer Graphics Metafile. A graphics file format format. Uses mainly vector graphics, but can also exchange raster (bitmapped) graphics.

Chad The little scrap of confetti that results from punching an old-fashioned machine-readable card, or from the holes in on the border of tractor-feed paper. A chadded form is one that has the holes punched clean through; a chadless form is one with the chads still hanging on by a thread.

Chaining A records storage process by which randomly stored records belonging to the same group are logically linked for retrieval.

Channel The one-way path by which signals are sent between devices.

Channel bank A multiplexer. A device which puts many slow speed voice or data conversations onto one high-speed link and controls the flow of those "conversations."

Channel map Specifies destination channels, output devices and patch maps for MIDI messages.

Character A single letter, digit or punctuation symbol. A character equals a byte.

Character-based interface A non-graphic, non-icon, I-miss-the-good-old-days, text-only user interface. Like DOS. Unlike Windows or MAC's OS.

Character cell One means of displaying text is to assign a matrix of dots, say 8 x 16 deep, to each character space. That's a character cell. Which dots are lit and which are dark determines the display of the letter. A character cell display is used in "chasing" lighted displays like the one on Times Square.

Character count A means of estimating the space text will take by averaging the number of characters in one line.

Character device A printer or other peripheral device that receives data character by character, instead of in bursty blocks of data.

Character generation. Software that adds type, for titles, credits, subtitles to videos.

Character locating reference mark Tiny dot placed dead center under each letter in a Linotronic's font. Allows the machine to place the character precisely within its em space.

Character Matching An OCR technique. The software contains "templates" of possible characters. When the scanner sees a letter, it compares it to its library of pattern templates. If it matches precisely, it translates it to the corresponding text character and sends the ASCII equivalent of the letter to the output file. A very accurate OCR method IF you can control the input. Not so good if you're scanning a variety of text — letters, articles, invoices — from many uncontrollable sources.

Character pitch The number of characters per inch. The "tightness" of letters a printer can accomplish.

Character recognition The ability of a machine to read human-readable text.

Character set The group of symbols, letters and/or codes that are used to represent a language. The character set for English has 26 characters. The ASCII character set has 128 characters.

Character terminal A terminal that cannot show graphics, only text.

Character triplet Three characters which are typeset as one unit for typographical purposes. See ligature.

Charge Coupled Device. See CCD.

Cheapernet Another name for thin Ethernet, a coax LAN that uses smaller-than-normal (and thus less expensive) cable. Same specs as Ethernet, except works on shorter distances.

Check bits In data transmission, the bits inserted for error-checking purposes. After a segment has been transmitted, a code corresponding to the segment is inserted, called the check bit. The receiving device accepts the data segment, and "checks" the check bit to see if they agree.

Check Mark Sense AKA Mark Sense. The ability of software to read check marks, Xs or filled-in boxes on a form, and populate a database with the appropriate data.

Checkpoint A disaster recovery scheme. A copy of new computer activity is made periodically. If there is a disaster, only data entered or changed since the last "checkpoint" is lost.

Checksum An error-checking scheme in data communications. When a packet of data is sent, the total of the ASCII character numbers represented by the data is sent along also. At the receiving end, the ASCII characters are totaled, and that sum is compared to the checksum. If they don't agree, there's been an error in transmission.

Child In database management, data that is dependent on other data is the other data's child.

Child program A second program which is executed by the main program you're running. The user is not required to instruct it to run. Simplifies the "user interface" of a complex program.

Chroma The level of saturation or intensity of a color.

Chroma key Method of electronically inserting the image from one video source into the picture from another video source using color for discrimination (see blue screen). A selected "key color" is replaced by the background image.

Chrominance The color portion of the video signal. Chrominance includes hue and saturation information but not brightness. Low chroma means the color picture looks pale or washed out; high chroma means the color is too intense, with a tendency to bleed into surrounding areas. Black, gray and white have a chrominance value of 0. Brightness is referred to as luminance.

Chooser Macintosh's utility that lets the user choose which printer, server, modem, scanner, etc., on the network he wants to use.

Cicero A European measure of line width or page depth. A cicero equals 0.178 inches.

CID Charge Injection Device. A way of making solid-state image sensors (CCDs.) The light-generated charge is sensed by injecting from the sensor into the substrate.

CIE Commission International de l'Eclairage. The international commission on illumination. Developer of color matching systems.

Cine-mode When human-readable data (e.g. microform) is recorded on a film strip so that it can be read when held vertically. Also called vertical mode. See comic-mode.

Click To press and release a mouse button quickly.

Click track A metronome-like timing signal that a control device monitors, to adjust the run speed accordingly. Performers, especially drummers, sometimes listen to a click track in headsets to keep the tempo steady. Purist musicians frown on click tracks, but today's reliance on sequencers and computers requires humans to be super-precise.

Client A computer that is configured to request services from a network server.

Client application The "destination" application in Windows and Windows NT that can accept linked or embedded objects.

Client/server computing Distributed network computing where applications are executed cooperatively by two computers: the client workstation, where processing is done; and the server, which holds and metes out files, databases, etc., as needed.

Climatizing Preparing paper to be used in printing devices — removing moisture, flattening, "fanning" etc. — to prevent jams.

Moore's Imaging Dictionary

AKA curling.

Clip A continuous set of frames from a source. A clip begins at one frame designated as the in point, and ends at one frame designated the out point. A clip can be thought of as a single scene or shot. A clip's frames can contain video only, both video and audio or audio only.

Clipboard The temporary cache of memory in a computer where data is held while it's being copied from one place to another. Used in Macintosh and Windows applications.

ClipBook page A Windows NT term. A piece of information you have pasted onto your local ClipBook. The piece of information (called a page) is permanently saved. Information on a ClipBook can be copied back onto the Clipboard and then pasted into documents. You can share ClipBook pages with others.

Clock speed The rate at which the computer clock oscillates, measured in MHz (one million cycles per second). The computer's clock rate determines its overall processing speed. Typical clock speeds for AT-compatibles are 6, 8, 10, 12, 16 and 20 MHz. For 80386 machines, typical speeds are 16, 20, 25, and 33 MHz. For 80486 machines, 33 and 50 MHz.

Clone 1. A raster image processor (RIP) that is not based on PostScript. Non-standard, but faster than PostScript, and may be OK for internal-only applications. 2. A PC that emulates an IBM computer.

Cluster 1. Unit of storage allocation used by MS-DOS usually consisting of four or more 512-byte sectors. 2. Group of terminals or workstations on the same system.

Clustering Storing documents that are related to one another near each other for faster retrieval. The idea is, if you want a certain document you are likely to seek a related one next.

CLUT Color Look-Up Table. The palette used in an indexed color system. Usually consists of 256 colors. The color table can be focused around a range of colors that predominate in your image, so a smaller bit depth (8 bits, 265 colors) can more accurately describe the image. For example, if your image is a red apple on a red tablecloth, you can choose mostly shades of red for your CLUT.

CLV Constant Linear Velocity. Data on an optical disc is recorded in a long continuous spiraling track (like a phonograph record with one long song). Used for videodiscs of movies, not for random access storage. The player adjusts the disc's spinning speed so that the larger outer tracks (which normally would spin faster) can be slowed down and thus hold more data than the smaller inner tracks. The speed drops from 1,800 rpm to 600 rpm.

C-mount Threaded lens mount first used in 16mm film cameras; now a closed-circuit and video camera standard.

CMYK Cyan, Magenta, Yellow and blacK. The four so-called process colors (technically, they are "subtractive" colors) that are used in four-color printed reproduction. In desktop publishing, it's one of the color models; the others being HSB, PMS and RGB.

Coated paper Printing stock that has a clear glossy finish for a more professional presentation. Why they call some magazines "slick."

Coax Slang for coaxial cable. Used in high-speed, high bandwidth data communications. See CATV.

COB Close Of Business. A deadline, usually.

Codec COder-DECoder. A device (actually a circuit chip) that converts analog input into digital form, and digital into analog. In video, the codec also compresses and decompresses the analog signal.

Coercivity The level of magnetism it takes to change the polarity of a bit in a magneto-optical write.

COLD Computer Output to Laser Disc. When "coded data" (ASCII databases, for example) generated by a host computer are stored on an optical disc. COLD replaces COM (Computer Output Microform) as a mainframe storage medium. Benefits: data can be removed from expensive mainframe storage, yet retrieved fairly quickly if need be.

Collaboration A multimedia term. Collaboration involves two or more people working together in real-time, or in a "store-and-forward" mode. Applications will enable a group of people to collaborate in real-time over the network using shared screens, shared whiteboards and video conferencing. Collaboration can range from two people reviewing a slide set on line to a conference of doctors at different locations sharing patient files and discussing treatment options.

Collapse To "hide" the subdirectories or folders-within-folders below a specified level.

Collate To arrange pages into the proper order.

Collision detection A network traffic control technique that detects if simultaneous interfering transmissions have taken place. If so, transmitting machine will wait a while and try again. Not too efficient. Collision detection is an essential part of the CSMA/CD access method.

Colophon Trademark and/or logo of the publisher, plus information about typefaces used, production methods, etc., printed on the last or title page of a book.

Color calibrators Souped-up densitometers which attach directly to the screen and read color values and intensities. They then adjust the screen to more closely match output colors, for truer color WYSIWYG. Calibrators manufactured by Radius, Minolta and others are relatively easy to use, and cost between $500 and $3,000.

Less expensive calibration utility programs analyze the performance values of specific monitors in great detail, and that information may be used to adjust to optimum performance. The main difference between a color calibrator and a calibration utility is that the calibrator allows for precise instrument adjustments, while the software must rely upon the user's eye to select the proper setting.

Color correction Any of a variety of digital or analog techniques for modifying the color of a video or image. Primary or overall color correction usually allows modification of the low luminance, midrange and high luminance areas of the picture. Controls allow adding and subtracting red, green, or blue (RGB) in these areas. Secondary or "spot" color correction allows one or many regions of the picture to be modified to different ranges of hue and saturation.

Color cycling A means of simulating motion in a video by changing the colors.

Color keying A trick motion picture and TV directors use to superimpose one image over another for special effects. An image (say a person behind the wheel of a fake car) is filmed against a solid color background (usually blue.) That film is sandwiched on top of another film sequence (say, the view from the rear of a moving vehicle going down the street). The result: the person appears to be driving a car down the street.

Color model How you describe a color. Imagine trying to explain "red" to a blind person. You can't, without a prearranged "language." Color models are those languages. See CMYK, HSB, PMS and RGB.

Color separation The process of dividing a full-color piece of artwork or photograph into its component process printing colors (CMYK).

Color shift The unwanted change of hue caused by too many reductions or alterations.

Color space The range of colors that can be expressed in a certain set of restrictions. For instance, the RGB color space comprises all the colors that can be created by mixing RGB.

Color suppression The technique in video that de-saturates the blue (usually, but can be any color) screen background so a chromakey effect can look more natural, and doesn't "bleed" over onto the subject.

Color temperature. High-res color monitors set their color temperature, or the intensity of light, at a very high 9,300 degrees Kelvin. This produces a more intense, attractive image. But the brighter light also shifts the colors toward blue. Color calibrating tools compensate for this mistake.

Column inch A text measurement one inch deep by the width of the column wide. Often how classified ads are sold.

COM Computer Output Microform. The process of converting data (having been input by a number of means) to microfilm or microfiche.

COM port. Same as serial port.

Comb In forms creation, a horizontal line with short vertical strokes in which hand-printed entries are to be made. Constrains handprinting for easier character recognition.

Comic-mode When human-readable data (e.g. microform) is recorded on a film strip so that it can be read when held horizontally. Also called horizontal mode. See cine-mode.

Command line The commands you type to run an application. In

DOS, the command line is the C: prompt. In Windows, it's in a dialog box.

Communication The exchange of information between two sources, in which none of the information is changed or deleted — at least not enough to make a difference.

Comp Short for comprehensive layout. A detailed full-size mock-up of a printed piece for approval.

Compact disc See CD.

Compiling The process of translating source code (what programmers write) into object code (what computers read).

Component video. A video signal where the red (R), green (G) and blue (B) picture components are recorded separately as individual signals. Synchronization information may be included with the G signal or be separate. On rare occasions luminance and chrominance signals (YIQ or YUV) may be provided as separate signals, also referred to as component signals. Separating the components results in better quality over many generations and eliminates cross-color and cross-luminance artifacts typical of NTSC and PAL composite video. Component video maintains higher color resolution for chroma keying, integrates without transcoding to most digital effects and graphics systems, and does not have color framing problems, all of which make it more suitable to high-end graphics and effect production.

Compose To prepare text into type,. i.e., typesetting.

Composite video The standard TV signal in which all the colors and the signaling (vertical and horizontal controls) are sent together. Unlike component RGB, in which the red, green and blue signals are separate.

Compositor Same as typesetter. Means either the machine or the person.

Compound document File that has more than one element (text, graphics, voice, video) mixed together. The elements were created from more than one application.

Compound term A multi-word phrase that can be treated in a retrieval system as one word. Example: "bill of lading."

Compound device Multimedia word. An MCI multimedia device that requires media files to operate. A MIDI sequencer is a compound device, since it needs a MIDI file. An audio CD player is a "simple," not compound, since it runs without outside software.

Compound document A compound document contains information created by using more than one application. It is a document often composed of a variety of data types and formats. Each data type is linked to the application that created it. A compound document might include audio, video, images, text, and graphics. Compound documents first became possible to PCs with the introduction of Windows 3.1, which included OLE (Object Linking and Embedding).

Compression A software or hardware process that "shrinks" images so they occupy less storage space, and can be transmitted faster and easier. Generally accomplished by removing the bits that define blank spaces and other redundant data, and replacing them with a smaller code that represents the removed bits.

Computer readable Data which is in a format, such as ASCII, or on a medium, such as disks, tapes, optical discs or punched cards, that a computer can understand. Same as machine readable.

Concantenation Putting individual strings of characters together into a single character string. Also, when voice synthesizing systems put separate words together, with the correct emphasis, to sound like speech.

Condensed Narrow type font. Used to pack more text into a small space.

Confidence A measurement of how certain an OCR is that it has correctly identified the character.

Conformance Means something meets standard and specifications. Conformance to a standard alone does not guarantee the products will work together.

Conservative recognition An OCR that uses rules and parameters that increases the chance of a correct identification. Results in a higher reject rate, but fewer mistakes. See aggressive recognition.

Constrained handprinting The ability of an OCR to recognized handprinted characters when the writer is restricted to certain rules

— must fit in a box or a "comb," must not cross or touch other letters, etc.

Contention What it's called when two or more users try to access the same device at the same time. There are "collision prevention" techniques to solve contention problems in LANs.

Contextual search To locate documents stored in a system by searching for text that appears in them, rather than by searching for them by file name or other indexing technique.

Contiguous Placed adjacently; one after another.

Continuity Matching all elements between shots in a video. For instance, if an actor is carrying a coffee cup in his right hand in one scene, you must make sure it's still in the right hand in the next scene. Continuity also means assuring that actions and speeds are continuous from scene to scene. Maintaining the illusion of coherent time and space within a project is the art of video editing.

Continuous tone An image that has all the values (0 to 100%) of gray (black & white) or color in it. A photograph is a continuous tone image. Contrast with halftone.

Contone Slang for continuous tone.

Contrast The degree of difference between the lightest and darkest tones in an image. Greater contrast means that the difference between black and white is greater, making the darker grays closer to black and the lighter grays closer to white. Less contrast there is less difference between black and white, resulting in more mid-level grays and fewer dark blacks and bright whites.

Control character A non-printing ASCII character which controls the flow of communications or a device. Control characters are entered from computer keyboards by holding down the Control key (marked CTRL on most keyboards) and pressing a letter.

Control menu A Windows/Windows NT term. A menu that contains commands you can use to manipulate a window. To open the Control menu, choose the Control-menu box at the left of the title bar in a window, or select an application icon at the bottom of the desktop. Every application that runs in a window and some non-Windows NT applications have a Control menu. Document windows, icons and some dialog boxes also have Control menus.

Control panel A window under the Apple menu that allows you to adjust various system parameters including the way the screen looks, the way the mouse works, sounds, and under System 7, memory attributes.

Control strips Series of color bars and percent tints placed just outside final image area; used to help maintain consistency during print runs.

Control track 1. A track on an optical disc that contains the formatting information necessary for writing, reading and erasing the data tracks. 2. The timing pulses recorded on videotape to maintain a constant playback.

Controller board or card An internal hardware add-on that fits in your PC's chassis and allows it to connect to and work with some external peripheral device, such as a printer or optical drive.

Convention The accepted way a thing is generally done. Syntax is a convention. My choice to spell acronyms in all caps with no periods was based on the accepted convention.

Conventional memory Up to the first 640K of memory in your computer. MS-DOS uses this memory to run applications. Beyond conventional you have extended memory.

Convergence In an RGB monitor, where red, green and blue signals all "converge" in one pixel. At full brightness, the RGB pixel in convergence would be white.

Coprocessor An additional central logic unit (processing chip) which performs certain dedicated tasks, freeing up the computer's main CPU.

Copy A duplicate of the original. A digital copy (from CD to CD, for instance) will be perfectly identical. The condition is binary; the signal is either on or off, no "noise" in between.

An analog copy (from mag tape to mag tape, for instance) will likely degrade each time a copy is made (called generations) because of tape noise.

Corona Part of a printer or copier that removes the previous image from the photoconductor to prepare for accepting the next image. Also known as a corotron.

Moore's Imaging Dictionary

Cover sheet Fax software feature that helps you prepare and automatically sends a cover page with each outgoing fax message.

CPI Characters Per Inch. The density of characters per inch on tape or paper. See pitch.

CPS Characters per second. A measurement of a printer's speed.

CPU Central Processor Unit. The computer's main chip on the motherboard which does most of the calculations and manages the operating system.

CRC Cyclical Redundancy Checking. An error-checking technique in data communications. A CRC character is generated at the transmission end. Its value depends on the hexadecimal value of the number of ones in the data block. The receiving end makes a similar calculation and compares its results with the sending machine's result. If there is a difference, the recipient requests retransmission.

Critical dimension The more important dimension between width and height. If you're printing a multi-page document, and you can print as many pages as you want, the critical dimension is the width (it's determined by the size of the paper.)

Crop To reduce the size of an image by trimming off unwanted parts. Sometimes cropping is combined with reduction to reduce the size.

Crossfade A gradual and smooth transition from one audio state (level) to another. You select the desired volume levels for the initial and final audio states, and the editing software controls the "ramp."

Cross-line storage A play on "on-line" and "off-line." refers to mixed media storage — combinations of magnetic disk and tape or optical products in a single subsystem.

Crosstalk Unwanted energy transferred from one circuit system to another, usually because they are close to each other.

CRT Cathode Ray Tube. The glass, vacuum display device found in television sets, computer terminals and oscilloscopes. At the back of the CRT is an electron gun which directs an electron beam to the front of the tube. The inside front of the tube has been coated with a material which reacts to and lights up once the electron beams hit. CRTs are very reliable if they are vented, since the electron gun gets hot.

CSMA (/CD, /CA) Carrier Sense Multiple Access. The most obvious problem with any resource sharing network, is "what happens if I want the printer at the same time someone else wants it?"

With CSMA, the stations "listen" (carrier sense) for any other station transmitting on the network. When it doesn't sense carrier, it sends. But collisions still occur. Two alternative versions (CSMA/CD and CSMA/CA) attempt to reduce both the number of collisions and the severity of their impact.

Carrier Sense Multiple Access with Collision Detection (CSMA/CD) does the same as above, but continues to listen during the transmission. If two stations try to get on the network at the same time, they both stop and retry after a random time out.

Carrier Sense Multiple Access with Collision Avoidance (CSMA/CA) does all the above, with the additional trick of using time division multiplexing (TDM), which allows different packets of data from different stations to travel one after the other, like cars entering a freeway.

CSU/DSU Channel Service Unit/Data Service Unit. A device that provides an interface between digital equipment and the (analog) network. Complies with FCC requirements and often includes capabilities to control switched networks and/or line conditioning capabilities. A fancy modem.

CTE Charge Transfer Efficiency. In a CCD, the percentage of charge that is successfully transferred from one CCD shift register element to the adjacent register.

CUI Character-based User Interface. Computer control system that makes the user type in commands (characters) to operate the computer. Opposite of GUI, which uses pictures, or "icons," to help the user operate the computer. PCs running MS-DOS use CUI; Macs use GUI.

Curie Point The temperature at which certain mixtures of elements (usually so-called "rare earth" elements and irons) relax their resistance to magnetic changes. In a magneto-optic drive, the surface to be marked is heated briefly by a laser light to its Curie point. Magnetism is then applied in the proper polarity to make the spot a "1" or a "0." It cools, and is locked in that position, until it reheated and changed again. This is how magneto-optic drives can be erasable.

Moore's Imaging Dictionary

Curling Preparing paper to be used in printing devices — removing moisture, flattening, "fanning" etc. — to prevent jams. AKA climatizing.

Cursive Type that looks like fancy handwriting. AKA script. Also used describe real handwriting, as against handprinting.

Cursor The symbol on a screen that shows where the next activity will take place. Graphics programs often change the shape of the cursor, depending on what action the computer is programmed to take next.

Cursor submarining A liquid crystal display on a computer laptop screen doesn't write to screen very fast. When you move a cursor across your screen or move your mouse quickly across the screen, the cursor disappears. This phenomenon is known as cursor submarining. Cute.

Curves and arcs Paint packages handle curves and arcs in a variety of ways. Examples include spline curves, wherein you specify a series of points and the package draws a curve that smoothly approaches those points, and "three point" bezier curves, in which the first two points anchor the ends of the curve and third selects the apex.

Cut 1. In video editing, a clean, abrupt change from one scene to another. The place where one video clip stops and another starts without any dissolves or fades. A quick transition. 2. To remove a graphic element of section of text in order to place it somewhere else.

Cut and paste To remove text or a graphic element from one document and placing it in another document. Usually accomplished by placing the cut element into a memory buffer (sometimes called a "clipboard") and then reading from the buffer for placement into the new document.

Cutaway In video editing, changing the scene to something other than the principal action. A typical cutaway: during a talk show, cutaway to the audience while the guest is speaking. Cutaways can maintain the continuity between two scenes. They also eliminate jump cuts. Also called an insert shot.

Cutline Same as caption.

Cuts-only Low-cost video editing systems only splice video clips together — no dissolves, wipes, fades or optical effects are available on cuts-only systems. Uses one source machine and one record machine.

Cyberspace The hugely romanticized notion that computer networks divulge some greater non-tangible worldspace, accessible to those with the right passwords and a 21st century brand of luck and pluck. Coined by William Gibson in his pretty-good sci-fi novel "Cyberpunk."

Cycolor A printing process that allows full color, full tonal reproductions of continuous tone images. Uses a special film that is embedded with microcapsules containing dyes.

Cyan One of the colored inks used in four-color printing. One of the subtractive process colors; reflects blue and green and absorbs red.

Moore's Imaging Dictionary

D The fourth letter of the alphabet. In ASCII, uppercase "D" is represented as hexadecimal 44; a lowercase "d" is hexadecimal 64. In EBCDIC, an uppercase "D" is hexadecimal C4; a lowercase "d" is hexadecimal 84.

D2 Composite digital videotape format for recording NTSC or PAL digital video on 19mm tape cassettes. The video is recorded in one 8-bit data stream. D2 is for high-end production, post-production and transmission where composite video is good enough. Can result in multi-generational artifacts.

D3 A Panasonic video recording standard which uses 1/2" tape in composite signal format.

D/A converter Digital to Analog Converter. Generic term for any device that changes digital (binary) pulses into continuous wave (analog) signals. Modems and codecs are D/A converters.

DAD Digital Audio Disc. Compact disc.

Daisy chain Connecting devices in a series. The computer's signals are passed through the chain from one device to the next. A SCSI adapter can support a daisy chain of up to seven devices.

Daisy wheel A printer that uses a rotating wheel with spokes. Mounted on the spokes are the various characters. The wheel rotates, bringing the character in line with the paper, then a hammer strikes the character against the ribbon, forming an image on the paper. Daisy wheel printers print "letter quality," but are slow.

Moore's Imaging Dictionary

DAL 1. Data Access Language. Apple software that allows Macintoshes to read databases on non-Apple computers.

2. DAT Auto Loader. Device that accepts a magazine of five or so DAT tapes, which are each addressable.

Dark current The signal generating from a photosensor — such as a CCD — placed in total darkness. Also known as dark noise.

Dark line image Same as positive image. The letters and characters are dark, and the background is light.

DASD Direct Access Storage Device. Any on-line data storage device. Usually refers to a magnetic disk drive, because optical drives and tape are considered too slow to be direct access devices. Pronounced DAZ-dee.

DASS Direct Access Secondary Storage. Same as near-line: storage on pretty-fast storage devices (e.g., rewritable optical) that are less expensive than hard drives but faster than off-line devices.

DAT Digital Audio Tape. A technology that records noise-free digital data on magnetic tape. Generally used for audio, a DAT cassette can hold up to 2 gigabytes when adapted for data storage.

Data According to AT&T Bell Labs: Data is "A representation of facts, concepts or instructions in a formalized manner, suitable for communication, interpretation or processing." I know data are supposed to be plural, but I hate the way "data are" sounds. Don't you?

Data communications The movement of data between points, including all the manual and machine operations necessary for this movement. Contrast with data transfer.

Data compression Reducing the amount of electronic "space" data takes up. Methods include replacing blank spaces with a character count, or replacing redundant data with shorter stand-in "codes." No matter how data is compressed, it must be decompressed before it can be used.

Data density The ratio of image to white space on a scanned page. Useful in estimating storage needs.

Data entry Using an I/O device (input/output device), such as a keyboard, to enter data into a computer. Normally thought of as manual key entry. But OCR and handprint recognition are making

55

automatic data entry systems possible.

Data frame A unit of information, represented as a page or two of data, including the margin around the frame.

Data header User-defined title for a data frame.

Data processing A generic term for business applications of computer systems for operational tasks — from bookkeeping to facilities management.

Data rate The speed of a data communications channel, measured in bits per second. See also baud.

Data set A collection of related data. Usually refers to the part of the data to be viewed, but can also include indexing information, commands, printing parameters, etc.

Data transfer The movement of data inside a computer system.

Data transfer rate The speed at which data transfers from a storage device to a computer, measured in kilobytes or megabytes per second.

Database Data that has been organized and structured in a disciplined fashion, so that access to information of interest is as quick as possible.

Database management programs form the foundation for most document storage indexing systems.

Database management system DBMS. Computer software used to create, store, retrieve, change, manipulate, sort, format and print the information in a database. Database management systems are probably the fastest growing part of the computer industry. Increasingly, databases are being organized so they can be accessible from places remote to the computer they're kept on.

Database object One of the components of a database: a table, view, index, procedure, trigger, column, default, or rule.

Database server Software that provides high-performance database access by splitting the DBMS function into a front-end component (where data is manipulated by users or applications) and a database-intelligent, back-end component (where data is stored, retrieved, and managed). In PC networks, the front-end component often resides on a PC controlled by a single user, while the back-end component resides on a high-performance PC that services requests for data submitted over the network by users.

dB Decibel. Measures gain or loss, usually referring to volume.

DBMS Database Management System.

DC Direct current. A form of power in which the current's polarity stays constant. Direct current is what comes from batteries.

DCR Document Composition Facility. The underlying commands in Standard Generalized Markup Language (SGML).

DCT Discrete Cosine Transform. The algorithm that prepares image data for lossy compression techniques, such as JPEG.

DDE Dynamic Data Exchange. Windows' method for moving data and operating instructions between applications. Here's how it might work: a user in a Windows personnel database application sees that a personnel record contains a scanned image of the employee. Clicking on a button in the record causes the database application to send a message to an image display application, retrieve the particular image, size it and display it. For this to work, the image application must understand the database message. There is no standard for these messages. Applications must agree on the language they use to communicate. Excel's command language is one method.

DDS Digital Data Storage. A DAT format for storing data. It is sequential — all data that is recorded to the tape falls after the previous block of data.

De facto standard A standard which is widely accepted, but has no official sanction. TIFF is a de facto standard. See de jure standard.

De jure standard. A standard which has gone through the accreditation process by an official governing body, such as ANSI or CCITT.

Dead sector In facsimile, the elapsed time between the end of scanning of one line and the start of scanning of the following line.

Deadly embrace Stalemate that occurs when two elements in a process are each waiting for the other to respond. For example, in a network, if one user is working on file A and needs file B to continue, but another user is working on file B and needs file A to continue, each one waits for the other. Both are temporarily locked out. The software must be able to deal with this.

Deblock To find individual sections of data (one record) from a larger block of data (the disk, the mag tape).

Decision point A "crossroads" in a workflow system, where some parameter, either automatic or manual, has to take place before the document or object can move to the next task.

Decimal Numeric representation based on 10, with 10 possibilities, 0 through 9, for each digit.

Decimation A basic problem with current imaging systems is the mismatch between capture resolution, typically 200 or 300 dpi for documents, and display resolution, typically 72 dpi on VGA monitors or 100 dpi on higher-end monitors.

To display images, every third or fourth pixel is dropped out, or "decimated." The result is a breaking appearance that is hard to read. Techniques to overcome this convert the pixels into blocks and correlate ratios of black to white dots with a gray scale. Since many monitors today support gray scale displays, the quality of rendition can be greatly improved. Unfortunately, this is very computation-intensive, and slow.

Decision point The point in a workflow where a document will continue on a divergent path, depending on criteria. This automatic decision point replaces manual decisions on where to send the work item next.

Decolumnization The process of reformatting multi-column scanned-in documents into a signal column. Generally, when you are processing a document for use in a word processing program, a single column of text is preferable to multiple columns.

Decompress To reverse the procedure conducted by compression software, and thereby return compressed data to its original size and condition.

Dedicated server A computer on a network that performs specialized tasks for the network uses, such as storing images. The word "dedicated" means that the computer is used exclusively as a server. It is not used as a workstation; no one sits in front of it. A dedicated server sits all alone, attached to its network, working happily all by itself.

Deep space Space at distances from the Earth approximately equal to or greater than the distance between the Earth and the Moon.

Deep Space Nine My current favorite television show. It's gotten better lately.

Default Preset conditions and attributes that determine the operation of a program or device. The default value is automatically assigned if you don't specifically request another one.

Default printer The printer that is used if you choose the Print command without first specifying which printer you want to use with an application. You can have only one default printer; it should be the printer you use most often.

Defect management Programs which control the power, focus or tracking of read or write devices in optical or magnetic recording systems to compensate for defects in the recording medium.

Deferred Processing Another term for batch processing. Using batch mode, you can quickly prescan your documents, capturing just the image of each page, then perform recognition on these images later, freeing your computer for interactive work.

Definition An expression of merit for image quality, but has no formal measurement. For video-type displays, it is normally expressed in terms of the smallest resolvable element of the reproduced image.

Degausser A device that removes unwanted magnetism from monitors or the heads in a tape or disk drive mechanism.

Deck A tape-based device for recording and playing back video in cassette or reel-to-reel format. Another word for VCR or VTR (video tape recorder).

Delimiter The "divider" character, often a comma, between separate fields in database records.

Delivery system In multimedia, the equipment on which the end user "plays" the presentation.

Delphi Consulting Group A great resource if you're thinking of implementing imaging. Even better if you're thinking of reengineering your workflow, based on imaging. 617-247-1511.

Delphi forecasting A questionable attempt to predict the future. Ask a group of experts what they think might happen, and then sort of average out their answers.

Demand publishing 1. The production of just the number of printed documents you need at the present time, as in "just in time." 2. The immediate production of printed documents which have been created and stored electronically.

De-migration The most common way to cache from optical storage is to prefetch images from an optical server and move them to magnetic media, either on a file server or at the workstation, before they are requested by the user. Some vendors call this "demigration" (migration is moving data to optical storage).

Denis the Little The 6th century monk who decided that history should be split between B.C. and A.D.

Densitometer Instrument that measures the density of printed images; often used to check uniformity of color in a press run.

Density Degree of darkness of an image. Also percent of screen used in an image. Also the number of bits (or bytes) in a defined length on a magnetic medium. Density describes the amount of data that can be stored.

Departmental imaging system A multi-workstation imaging system, typically used by at least three and usually by more than 25 people in a workgroup or department. See desktop imaging system.

Depth of field The range of distances away from the scanner within which the image stays in focus. Most scanners assume the target document is perfectly flat, therefore their depths of field are very narrow. A wide depth of field is required for scanning books or periodicals that do not lay flat. Bar code scanners have a broad depth of field, since the distance from the scanner to the coded object usually varies a lot.

De-rated The real-world performance of something, as against the manufacturer's rated performance of the product.

Descenders A typographic term for the portion of lowercase characters that falls below the main body of the letter. The lowercase letters g, j, p. q and y have descenders.

Descending sort Sorting data "backwards," from high to low, or from Z to A.

Descriptor The key word, code or phrase that an automated document retrieval system uses to identify and locate the document. Descriptors sometimes "summarize" the most relevant data in the document, so that reading the descriptors — rather than retrieving the entire document — is sometimes sufficient for the purposes of the search.

Deskewing Adjusting — straightening — an image in software to compensate for a crooked scan.

Desk accessory A Macintosh features that allows certain programs to be available no matter what application is running. Analogous to the "path" in MS-DOS applications.

Desktop 1. Slang for any computer function that can be done on a standalone PC, rather than a larger, more powerful, computer.

2. The background of your Windows and NT screen, on which windows, icons, and dialog boxes appear.

Desktop imaging system An imaging system with a single workstation — often a microcomputer — meant to be used by only one person at a time.

Desktop pattern A design that appears across your Windows desktop. You can create your own pattern or select a pattern provided by Windows or NT.

Desktop publishing The term applied to the creation of printed documents using a PC. The documents may be printed directly from the desktop publishing application software (usually with a desktop laser printer), or prepared for a commercial printing process. Do not confuse with "electronic publishing," which refers to electronically preparing documents which are distributed and read electronically.

Desktop video Using non-dedicated PCs and software to do video editing and post-production tasks.

Deselect Deselecting an item means the object or program is no longer ready to manipulate or run.

Destination document The document into which an object is linked or embedded via OLE.

Developer Person or company who develops new software, or customizes existing applications, to do specific things for you or your company.

Developing agent The chemical in a photographic developer that converts exposed silver halide to visible black metallic silver.

Device contention The way Windows NT allocates access to peripheral devices, such as a modem or a printer, when more than one application is trying to use the same device.

Device driver Small program that tells the computer how to communicate with particular types of peripheral devices. You may have a certain device connected to your system, but your operating system won't "recognize" it until you have installed and configured the appropriate driver. In Windows and NT, device drivers usually have a .DRV extension.

Device driver interface A function in a API that allows developers to write drivers that write directly to hardware and do not rely on interrupts.

DIA/DCA Document Interchange Architecture. IBM-endorsed architectures, part of SNA, for transmission and storage of text, data, voice or video documents over networks. Becoming industry standards by default.

Diacritical mark A symbol placed above or below a character in non-English languages that denote a certain pronunciation. Examples are accent marks and umlauts.

Dialog box A temporary window which prompts you to input information or make selections necessary for a task to continue. Some dialog boxes present warnings or explain why a command cannot be completed.

Diazo film In micrographics, the type of film used to create same-polarity copies (i.e., negative image of negative-reading original). Diazo film is exposed with the original under ultraviolet light, then processed with either aqueous (liquid) or anhydrous (gas) ammonia.

Diazonium salt The light-sensitive component of diazo and vesicular films.

DIBACOS Dictionary-Based Compression System. Text compression technique based on the premise that 2/3 of the words in any document are among the 1,024 most commonly used in the language. DIBACOS algorithms compress text by 60%.

Didot point A European type measurement. Equals 0.0148 inches.

Dielectric process A printing technique where special conductive paper is exposed to tiny electrically charged needles.

Differential spacing In typography, allowing letters to take up varying horizontal space in relation to their widths. For example, an "i" takes up less space than an upper-case "W". Opposite of fixed spacing, where each letter is assigned the same space, regardless of its shape or width.

Digit A character that represents a number, not a letter. 0 — 9 are digits. In a hexadecimal counting system, letters are sometimes digits, too. Confusing.

Digital The use of binary code to record information. "Information" can be text in a binary code like ASCII, or scanned images in a bit mapped form, or sound in a sampled digital form, or video. Recording information digitally has many advantages over its analog counterpart, mainly ease in manipulation and accuracy in transmission.

Digital audio The storage and processing of audio signals digitally. It usually requires at least 16 bits of linear coding to represent each digital sample.

Digital camera The newest generation of video cameras transform visual information (lightness and darkness) into pixels, then translate the pixel's level of light into a number (or, in the case of color, into three numbers — one for the level of red, green and clue in the pixel).

These digital images can then be manipulated pixel by pixel to create exciting new applications in video and film production. They can also be compressed, stored and transmitted in more or less the same manner as traditional digital data.

Digital monitor Receives discrete binary signals to paint the screen. Monitors generally were of this type before VGA models appeared. Digital monitors do not have as wide a range of color choices as analog types; digital EGA monitors, for example, can display just 16 colors out of a palette of 64.

Digital typesetter Type creation device that stores fonts in digital format rather than on photographic negatives.

Digital video A video signal represented by machine-readable binary numbers that describe a finite set of colors and luminance levels.

Digital video tape recorder DVTR. Any of several VTRs from Sony or Panasonic that records using the D1, D2 or D3 recording format. Any VTR that records video and audio in any digital recording format.

Digitize To convert an image or signal into binary code. Visual images are digitized by scanning them and assigning a binary code to the resulting vector or raster graphics data. Sounds are digitized by recording frequent "samples" of the analog wave, and translating that data into binary code. "Digitizing" is often used interchangably with "scanning."

Digitizer Strictly speaking, any device that converts analog data to digital. But generally reserved to refer to digitizing tablets.

Dimensioning The standards, applied in CAD programs, that govern drafting.

Dimmed How a GUI indicates that a feature or action is unavailable or disabled. A dimmed command is displayed in light gray instead of black, and it cannot be chosen. It shows that the command is sometimes available, just not now.

Dingbat A typographical decoration. Used to mark the end of stories, or to tart up a page of text. Also called printer's marks. Some dingbats:

● ☞ ✓ ✪ ✚ ✸ ✂ ✝ ✎

DIMS Document Image Management System. Marginally unfortunate acronym, but a lot of people use it.

DIP 1. Document Image Processing. Dumb acronym. Nobody uses it. 2. Dual In-line Package. A common integrated circuit with parallel rows of pins which serve as connectors. The pins can be switched on or off by the user — these are called DIP switches.

Dipthong In typesetting, two vowels which are joined to form a single character, also known as digraph. A special form of ligature.

Direct access method A data search technique that uses an identifying word or code in the document itself as its address.

Direct color This means the number of colors in the system palette is equal to the number of colors created by scanning in the image. This is not efficient: even though an image is scanned at 24 bits, it is not necessary to use 16.8 million colors to reproduce it. Index color is more efficient.

Direct image film Film that maintains the same polarity — positive for positive, negative for negative — as the image which it is duplicating.

Directory Part of a structure for organizing your files on a disk. A directory can contain files and other directories (called subdirectories). The structure of directories and subdirectories on a disk is called a directory tree.

Directory tree A graphical display of a disk's directory structure. The directories on the disk are shown as a branching structure. the top-level directory is the root directory.

Dirty power Alternating current (AC) that is not a perfect sine wave and not perfectly 120 volts. Electricity can be made "dirty" by: spikes or glitches (transient impulses of high amplitude but very short duration); sags and surges. Power can also be delivered consistently beyond its rating. In New York, Con Edison guarantees 120 volts plus or minus 10%. Any level below 108 volts or above 132 volts Con Edison would consider dirty power.

Disc Same as optical disc. A digital storage medium. Optical discs are made of a metal alloy recording surface sandwiched between a rigid substrate and a plastic protective coating. Lasers record data in the metal alloy by either creating tiny pits (ablation technique) or by causing small bubbles to form in the "negative" area, thereby reflecting the laser away.

In this dictionary and in IMAGING MAGAZINE, the rule is: disc with a "c" means optical disc. Disk with a "k" means magnetic hard or floppy disk.

Discretionary hyphen A hyphen that is manually typed into a word in the place of your choice, but doesn't appear unless your spot coincides with a line break. Overrules the place where the dictionary wants to hyphenate a word.

Disk Same as magnetic disk. A round, flat magnetic recording medium with one or more layers deposited on the surface which data can be recorded onto.

Disk array Combining rodundant disk drives for more capacity, or for disaster recovery. See RAID.

Disk cache Place in memory where frequently recalled data is stored while you're working. It makes retrievals much faster.

Disk controller A hardware device that controls how data is written to and retrieved from the disk drive. The disk controller sends signals to the disk drive's logic board to regulate the movement of the head as it reads data from or writes data from or writes data to the disk.

Disk drive A device containing motors, electronics and other gadgetry for storing (writing) and retrieving (reading) data on a disk. A hard disk drive is one which is generally not removable from the machine. A floppy disk drive accepts removable disk cartridges.

Disk duplexing A method of failsafe protection, occasionally used on file servers on local area networks. Disk duplexing involves copying data onto two hard disks simultaneously, each through a separate disk channel. The idea is, if one disk or channel is faulty, the other will most likely continue to operate normally.

Disk management Refers to the control of information stored on a disk. The logical relationship of subdirectories to root directories, for instance.

Disk mirroring A fault-tolerant technique that writes data simultaneously to two hard disks using the same hard disk controller. The disks operate in tandem, constantly storing and updating the same files. Mirroring alone does not ensure data protection. If both hard disks fail at the same time, you will still lose data. See disk duplexing.

Disk pack A cartridge of hard disk platters arranged as a single unit. A disk pack contains more space for storing and retrieving information than one single disk. See disk array.

Disk sector Magnetic disks are typically divided into tracks, each of which contains a number of sectors. A sector typically contains a predetermined amount of data, such as 256 bytes.

Disk (file) server A mass storage device that can be accessed by several computers, usually through a local area network (LAN).

Disk striping Spreading data over multiple disk drives. Data is interleaved by bytes or by sectors across the drives.

Diskette A floppy disk.

Diskless workstation A computer with processing power but no disk storage device. Diskless workstations are usually arranged on a local area network, and share a disk server.

Display Another word for monitor. The computer's screen.

Display attributes The way your monitor represents basic things — the background and reverse video colors, number of characters per line, number of lines, usually adjustable.

Dissolve The tail of one video clip blending or mixing into the head of another. Fade-outs and fade-ins are examples of dissolves to and from black, white or another color or texture. Dissolves between two video clips requires two source VTRs in an A/B-Roll configuration.

Distribution microform A duplicate copy of a microfilm or microfiche that is intended to be used in a viewer. Distinct from the master or intermediate microform, from which distribution copies are made.

Dithering A means of simulating gray tones by altering the size, arrangement or shape of background dots.

DLL Dynamic Link Libraries. A library is a set of software routines that provide services. These services are then used by other pieces of software. In the old days, libraries were compiled into an application before it could be run. Windows brought us Dynamic Link Libraries, which can be used on the fly, without being compiled into the application. If an application is built using DLLs, these can be updated without having to recompile the entire application. This makes it easier to upgrade. It also makes it possible for third parties to contribute to an application by providing DLLs. This expands the possibilities and lifecycle of an application.

D-max The highest density (the darkest image) possible for a particular photographic medium.

D-min The minimum density)the lightest image) possible for a particular photographic medium.

Document A collection of data, organized into some logical order and placed in a file or on paper.

Document Caching An "intelligent" way to speed up jukebox retrievals. It is based on three principles: (1) data is retrieved in more or less predictable ways; (2) no data access is truly random; and (3) the probability of re-access is higher than the probability of initial access. Caching makes an educated guess and retrieves documents you haven't asked for yet. These are staged in your local PC or in a network server, so when you DO ask for them, they are available instantly.

Document management The capture, indexing and intelligent retrieval of digitized (scanned) and electronic (wordprocessed) documents, as well as computer applications files (spreadsheets). Much excellent document management software and hardware is available, in wildly ranging prices and quality levels, for single users all the way up to enterprise-wide computer networks. Imaging Magazine specializes in writing about these products and the hardware that supports them. Call 1-800-677-3435 to subscribe ($17.95 a year).

Document mark Same as blip. An indicating mark exposed on microfilm for counting or timing purposes.

Document recognition The ability to capture all the information on a page (text and images) and recognize not only characters, but page structure (e.g., number of columns) and images and artwork.

Document retrieval The ability to search for, select and display a document or its facsimile from storage.

Document staging In a document retrieval from an optical jukebox, the process where the image is fetched from the server by the software, and stored on the user's local PC until it is used. After it is used, it has to be manually deleted from the PC's disk (the "original" of the image is still in the jukebox), leaving magnetic disk space for other documents to be staged.

Documentation Written text describing the system, how it works and how to work it. The manual.

DOD Drop on demand A type of inkjet printer.

Moore's Imaging Dictionary

Domain The small area that contains magnetism on a magneto-optical disc.

Dongle A device to prevent copies made of software programs. A dongle is a small device supplied with software that plugs into a computer port. The software interrogates the device's serial number during execution to verify its presence. If it's not there, the software won't work. Also called a hardware key.

DOS Disk operating system. Controls the basic functions of your computer. In usage, DOS usually means MS-DOS.

Dot Usually refers to halftone dots, and measured as a fraction of an inch — e.g. 133 dots per inch. Also refers to the basic element of resolution of a laser printer — e.g. 300 dots per inch. Do not confuse output resolution (measured in dots per inch) with scanner or monitor resolution (measured in pixels per inch).

Dot gain The tendency of ink to spread out on paper during printing. Bad for detail.

Dot leaders A series of dots that fill in the space between two items on the same line that help the eye follow across the page. Often used in tables of contents.

Dot matrix Type of printer that produces text and graphics from many small dots arranged in a matrix.

Dot pitch The distance from the center of one phosphor dot in a CRT to the center of the nearest phosphor dot of the same color on the adjacent line. The smaller the distance, the sharper the monitor. Dot pitch is the major determinant in the clarity of an image on screen. And you can't do anything about it. When you buy a monitor, you buy it with a certain dot pitch and you're stuck with that dot pitch.

Double buffering Using two buffers to temporarily hold data being moved to and from an I/O device. Double buffering increases data transfer speed because one buffer can be filled while the other is being emptied.

Double-click Clicking the mouse button twice rapidly while the cursor is over an icon. Usually launches a program or displays an image.

Downward compatibility The ability to read and write earlier generation or smaller-sized media, such as floppy disks. For example, Syquest makes an 88-meg removable hard disk drive that is "downward compatible" with — i.e., can read — its 44-meg disks.

Downloaded fonts Fonts that you send to your printer either before or during the printing of your documents. When you send a font to your printer, it is stored in printer memory until it is needed for printing. Some PostScript printers do not support downloaded fonts. They use the fonts that are permanently stored in their read-only memory.

Downsizing When companies move from large computer systems to smaller ones. There are four major reasons companies downsize from mainframe-based computer to local area network-based computing: 1. They save money. 2. Servers today have the power of mainframes 10 years ago. Some have more. 3. Servers are manufactured from off-the-shelf, readily available standard components. There is constant competition and improvement in quality and features. 4. Servers are much more flexible tools to design networks. You can start with one baby network containing one server and several workstations (aka clients) and confined to one floor of one small building and grow to a huge, complex network containing thousands of workstations, dozens of servers and spanning the globe.

DP The lowest quality, but fastest, setting on some printers. Stands for Data Processing mode.

DPI Dots Per Inch. 1. A measurement of output device resolution and quality. Measures the number of dots a printer can print per inch both horizontally and vertically. A 600 dpi printer can print 360,000 (600 by 600) dots on one square inch of paper. 2. A measurement of scanner resolution. The number of pixels a scanner can physically distinguish in each vertical and horizontal inch of an original image.

Draft quality A lower quality, but faster, setting on some printers. Useful for quick reviews, before committing to LQ — Letter Quality.

Drag 1. Moving an item or window on the screen by locating the cursor on the object, pressing the mouse button down and move the mouse across your desktop. When you release the mouse button, the window will remain in its new location. Apply this technique to drag any data object, such as icons or list box items. 2. A touchscreen term, where your finger does basically what a mouse does: selects and moves the cursor point across the screen.

Drag and drop Moving one object from a desktop application with a mouse, across the desktop and placing it ("dropping it") on another application. Most of the graphics operating systems, like Windows, Apple's Macintosh and Sun Sparc use Drag and Drop.

DRAM Dynamic Random Access Memory, the kind of RAM used in SIMMS. The data stored in Dynamic RAM (as opposed to "Static RAM") fades away over time, so the information has to be refreshed (rewritten) every few thousandths of a second. Although this consumes much more power than the static design, each memory cell can be built from just one transistor instead of the four required for static RAM so higher memory capacities can be built economically.

Draw Direct Read After Write. An error correction method that checks immediately to see if the data was successfully transferred. It's slow but reliable.

Drawing tools Used in painting and drawing software to create freehand lines or basic geometric shapes. Paint packages often provide an ellipse-drawing function as a variation of the circle (or vice versa) and a square drawing function as a variation of the rectangle. Virtually all packages offer filled geometric figures, the fill being either a solid color or a pattern.

Driver software A driver (which is always software) provides instructions for reformatting or interpreting software commands for transfer to and from peripheral devices and the central processor unit (CPU). Many printed circuit boards which you drop into a PC require a software driver in order for the other parts of the computer and the software you're running to work correctly. In other words, the driver is a software module that "drives" the data out of a specific hardware port. The port in question will usually have another device connected, such as a printer or modem, and the driver will be organized in software (i.e. configured) to communicate with the device.

Drop-frame For NTSC video, time code is normally produced by a generator which counts at 30 frames per second. However, NTSC color signals actually have a frequency closer to 29.97 frames per second.

Drop-frame time code compensates for this time difference by eliminating a certain number of frames from the time code according to an established formula so that it matches clock time.

Moore's Imaging Dictionary

Drop out 1. A section of magnetic media with a defect, on which the writing device is unable to record data. 2. Text or artwork printed in an ink that a scanner or copy camera can't see. Used to print text that a human should see, but can be left out of a scan. Forms processing sometimes uses drop-out ink to print the form itself, which isn't visible to the scanner, and hence leaving only the important data behind.

Drum scanner An ultra-high resolution scanner, often used for scanning transparencies. The slide is attached to a rapidly spinning drum, which turns under the scan head. It moves very slowly across the image, resulting in very close lines of dots.

Dry Audio recording with no electronic effects.

Dry imager The powdery toner used in electrostatic printers and copy machines.

Dry processing Method in which an exposed latent image is made visible without chemical treatment (usually a heat process).

Dry silver film Thermal process film. A non-gelatin silver film type which is processed by heat, not chemicals.

DSP 1. Digital signal processing. Transforming an analog signal into digital and processing numerically. The numeric processing part is key; you may filter or transform an image to enhance or recognize certain features with DSP.

2. Document Storage Processor. A computer that manages the storage devices of an imaging system. Database transactions are managed by the DSP.

DSR Document storage and retrieval. I don't know anyone who uses this acronym. But now you know what it stands for, in case someone should.

DSU/CSU Data Service Units/Channel Service Units. The devices used to access digital data channels. At the customers's end of the telephone connection, these devices perform much the same function for digital circuits that modems provide for analog connections. For example, DSU/CSUs take data from terminals and computers, encode it and transmit it down the link. At the receive end, another DSU/CSU equalizes the received signal, filters it and decodes it for interpretation by the end-user.

DTMF Dual Tone Multi-Frequency. A fancy term for push button or Touchtone dialing. (Touchtone is a registered trademark of AT&T, hence the capital "T").

DTP Desktop publishing.

Dummy A full-size mock-up of a layout or a printing project. Usually has fake "greek" text and merely representative images.

Dump Printout of a file's contents without any report formatting.

Duplex The ability of a scanner to scan both side of a sheet simultaneously. Requires two scanner heads and often two processing boards.

Duplex paper Paper that has a different coating or surface treatment on each side.

DVI 1. Digital Video Interactive. An Intel product. Not a compression technique per se, but a brand name for a set of processor chips that Intel is developing to compress video onto disk and to decompress it for playback in real time at the NTSC standard video rate of 30 frames per second. The chip set includes both a pixel processor, which performs most of the decompression and also handles special video effects, and a display processor, which performs the rest of the decompression and produces the video output. DVI's greatest long-term advantage, according to Nick Arnett writing in PC Magazine, is that its microprocessors are programmable, so DVI can be adapted to a variety of compression and decompression schemes. The DVI capacity is 72 minutes of uncompressed video on a single CD-ROM.

2. Device independent. A file that has been formatted in such a way that it can be translated into a fom that can be printed by any type of printer.

Dvorak keyboard A keyboard, invented mainly by August Dvorak, on which letters and characters are arranged for faster and easier typing than on the standard QWERTY keyboard. The QWERTY keyboard was actually designed to be difficult to use, to slow down typists so they wouldn't jam the old typewriters' mechanisms!

DVTR Digital Video Tape Recorder. Any VTR that records in D1, D2 or D3 digital recording format.

Dye polymer recording A thin layer of plastic, sandwiched in an optical disc, is heated by a laser to form either a bubble or a pit. The affected areas can represent "ones," the non-lasered areas can be the "zeroes." Dye-polymer is at present a WORM technology. It can be rewritable (a second laser can "relax" the bubbles), but there are no commercially available products.

Dye sublimation Printing process similar to thermal transfer, where exactly measured temperatures control the amount of ink transferred from colored ribbons to paper. Under high temperature and pressures, the ink is not melted, but is transformed directly to gas, which hardens on the paper after passing through a porous coating. Dye sub printers create very nearly continuous tones, making them great for natural images. Because the gas makes "fuzzy" dots, dye sub is not recommended for sharp-edged "computer-y" graphics or type.

Dylux proof Type of printer's proof made by exposing a photosensitive paper through a printing negative. Also called blueline.

Dynamic beam focusing When you have a curved cathode ray tube, the distance between the gun which shoots the electrons and all the parts of the screen are equal. When you have a flat screen, the distance varies slightly. Some beams have to travel further. Some have to travel not so far. Dynamic beam focusing focuses each electron to the precise distance it must travel, thus ensuring edge-to-edge clarity on the screen.

Dynamic Data Exchange See DDE.

Dynamic information In OCR, specific knowledge sources for the particular page or documetn — special characters, thesaurus words, etc. See Static information.

Dynamic Link Library See DLL.

Moore's Imaging Dictionary

E The fifth letter of the alphabet. In ASCII, uppercase "E" is represented as hexadecimal 45; a lowercase "e" is hexidecimal 65. In EBCDIC, an uppercase "E" is hexadecimal C5; a lowercase "e" is hexadecimal 85.

EBCDIC Extended Binary Coded Decimal Interchange Code. An 8-level code, like ASCII, developed by IBM. Supported by some OCR programs.

ECC Error Correction Code. In the event of read-write error, ECC is a method of recovering a block — 2,048 bytes — of data, applied by CD-ROM drives and in the mastering of CD-ROM discs.

Echo check Error control technique where commands typed to a remote computer sends the message back to the sender for verification.

Echoplex Error control technique where the commands typed into the local computer to a remote computer are returned, via a full-duplex link, to the local computer while you type.

ECN Engineering Change Notification. When the designers improve some part of a product that is already being manufactured, it often takes weeks or months before the entire manufacturing process — raw material ordering, personnel training, warehousing — catches on. That loses money. ECN systems use document management linked with CAD systems to manage that change notification.

EDAC Error Detection And Correction. An error detection scheme. It works this way: An prearranged extra block of data is added to each block written. After writing, the extra data is read back. If the extra data is correct, EDAC assumes the entire writing procedure went smoothly, and goes on to repeat the procedure with the next block. If an error is detected, the write process is repeated.

Edge connector Strips of brass or other conductive metal found at the edge of a printed circuit board. The connector plus into a socket of another circuit board to exchange electronic signals.

EDI Electronic Data Interchange. Automatic paperless system that is supposed to allow vendors to exchange invoices, purchase orders, etc., in a standard format. Standards are set by the X12 committee.

Edit Modify data, text and images after their initial creation. Image editing software is often called image manipulation.

Edit codes Codes inserted during formatting for later searching purposes.

Edit master The final videotape resulting from electronic editing; the videotape resulting from editing material from one or more source tapes onto a record tape. The edited master is at least one, and often two or three generations removed from the original source footage. Other distribution copies will be recorded from this edit master.

Edit suite An old word for the room in which video editing is performed. So named because the video edit room was divided into two rooms (a suite) — one holding the rack equipment and the second holding the remote control editing devices.

Editable PostScript PostScript commands that have been translated into a text file, which can then be changed without the need to use the applications program from which the PostScript file was originally created.

Editor A text editing program, used for altering programs, that lacks the formatting and presentation features of a full-featured wordprocessor. When you buy MS-DOS, one of the programs you get is a rudimentary editing program called EDLIN. Most programmers, however, now prefer more advanced editors, like ZEdit. This dictionary was written in ZEdit, a derivative of the excellent QEdit available from Semware in Marietta, GA.

EDDP Electronic Document and Printing Professional. A professional in the electronic document systems industry.

EDL Edit Decision List. In video editing, a series of time-code numbers that describe a series of video clips, cuts, edits and special effects. It controls the source deck(s), compiles all the source material, lines up the clips and records everything to the recorder deck. The EDL is a text file that describes how an edit should take place. Sort of like PostScript, but for video. A service bureau can use your EDL to output high-quality video.

EDMICS Engineering Data Management Information and Control System. The Navy's imaging initiative.

EDMS Engineering Document Management System.

Edutainment Unpopular answer to the question "What do you get when you cross educational material with interactive video?" A term coined by "someone who obviously knows nothing about either education or entertainment," says Laura Buddine, president of multimedia games maker Tiger Media.

Effective reduction The number of times a hypothetical document would have to be reduced to equal the size of a COM-generated microimage.

EGA Enhanced Graphics Adapter. A display technology for the IBM PC. 16 colors at 640 x 350 resolution. It's been replaced by VGA.

Egyptian Style of font with square serifs and heavy characters with consistent widths.

EISA Extended Industry Standard Architecture, a 32-bit bus standard introduced in 1988 by a consortium made up of AST Research, Compaq, Epson, Hewlett-Packard, NEC, Olivetti, Tandy, Wyse and Zenith.

Electron gun The device in the CRT that produces the electron beam that activates the phosphors, causing them to emit red, green and blue light.

Electronic composition Setting type and layout out pages with computer-based publishing systems.

Electronic forms Graphics that are merged electronically with data to appear one each frame. Can be as simple as a border box, or a logo or running header.

Electronic messaging Messages sent and received through an electronic medium in digital form. Electronic mail (e-mail) can be delivered via wire or wireless networks; it can be on a local area network (LAN) or a wide area network (WAN); it can be domestic or international.

Electronic overlay A form template that you merge with raw data to make it easier to read. The form will not remain part of the data; it is "overlaid."

Electronic publishing Producing and providing documents in electronic form. Do not confuse with publishing electronically (desktop publishing.)

Electrophotographic printing The technology used in copy machines and laser printers. An electrically charges drum is hit with small beams of light. Wherever the light hits, the drum loses its electrical charge. When toner is applied, it sticks to the non-charged parts of the drum. Paper is then pressed against the drum, and the toner adheres to the paper. The paper is then heated to "set" the toner.

Electrostatic printing Printing process that uses a special paper which is charged by an electron beam. The tone sticks to the charged areas. Used in large-image plotters.

Element printer Any printer that requires an interchangeable device for fonts that it strikes against the paper. An IBM Selectric, with its "ball," is an element printer. A daisywheel printer is also an element type.

Elevator seek or sort A logical way of organizing many requests for data from a storage device, especially a jukebox, in order to speed up processing. Incoming requests are sorted by priority depending on the data's location on the disc, or off-line. It responds to requests for data nearest the read head first, regardless of the order in which the requests arrived. Think of an elevator: It stops on the floors in the order it passes them, not in the order the riders pushed the buttons.

Elite A font that fits 12 characters in a horizontal inch.

Ellipsis Three equally spaced periods, used to indicate omitted or missing material, especially in quoted text.

Em A relative measurement of horizontal space — it's a measurement because it is equal to the width of a capital "M". It's "relative" because it's the width of the capital "M" in whatever font and size you're dealing with.

Embed To insert an object (file, image, sound clip) created in one document into another document (most often the two documents were created with different applications). The embedded object can be edited directly from within the document. The basis of OLE — Object Linking and Embedding.

Embedded object Information created in one document and inserted into another document (most often the two documents were created with different applications). Embedded objects can be edited from within the destination document.

Embedded SCSI A hard disk that has a SCSI (Small Computer System Interface) and a hard disk controller built into the hard disk unit. See also SCSI.

Embedded SQL SQL statements embedded within a source program and prepared before the program is executed.

Embedded system processors National Semiconductor's line of high-performance microprocessors used in dedicated systems, such as fax machines and laser printers.

Emoticon This is fun. From Emotional Icon, one of a growing number of typographical cartoons used on BBSs (Bulletin Board Systems) to portray the mood of the sender, or indicate physical appearance. Turn this book sideways with the left edge at the top.

:-D   writer talks too much

:-#   writer's lips are sealed

:-o   writer is surprised

:-&   writer is tongue-tied

:)   is a smiley face

;)   is a smile with a wink

;(is a frown with a wink

(:(is very sad

;?   is a bad guy

EMS Expanded Memory Specification. Several years ago, three computer companies — Lotus, Intel and Microsoft — jointly developed EMS. This standard defines how an MS-DOS program can access memory beyond 640KB while running under MS-DOS. Applications that conform to EMS (sometimes called LIM-EMS for Lotus/Intel/Microsoft Expanded Memory Specification) can take advantage of the computer's memory beyond 640KB of RAM. LIM-EMS uses a portion of the reserved memory area (between 640KB and 1MB) to access RAM beyond 1MB. Software that supports expanded memory uses this window to pass pages of data to and from expanded RAM as needed.

Emulsion The light-sensitive coating on photographic media, usually silver salts suspended in gelatin.

En Half the width of an em.

Enable To "turn on" a feature that was suppressed. Also, an integral technology upon which another technology depends. Compression is an enabling technology for fax.

Encapsulated PostScript An image description format. EPS translates graphics and text into descriptions to a printer of how to draw them. The font and pictures themselves need not be loaded into the printer; they've been "encapsulated" into the EPS code. The EPS file holds all the document data as un-rendered graphics and PostScript font mathematical data. Final rendering of the graphics file occurs in the laser printer or imagesetter, and is optimized to take advantage of the particular printer.

Encapsulation In object-oriented programming, the grouping of data and the code that manipulates it into a single object. If a change is made to an object class, all instances of that class (that is, all objects) are changed. Encapsulation is one of the benefits of object-oriented programming.

Encoding The basis of image compression. Means replacing strings of black or white pixels with (shorter) code numbers that describe them and their relationship to each other.

End-of-page A control character which stops the printer when it reaches the bottom of a file. Allows the operator to change paper, or adjust settings.

End-use microfilm Same as distribution microfilm; a copy that is meant to be used in a reader.

Engine A vague term for the hidden proprietary technology that actually does all the work. A document management software usually has an OCR engine of one brand or another.

Enhanced Small Device Interface (ESDI) A standard for magnetic storage drives, letting them communicate at high speed with a computer.

Enterprise A business that has one or more information networks and needs to apply consistent management policies and procedures.

Enterprise engineering Developing a corporate-wide information systems strategy and resource pool. Ensures all information strategies — hardware, systems, applications — all contribute to the enterprise's long- and short-term goals. What it often means in real life is a company has bought a bunch of incompatible stuff with no central vision or guidance, and is suffering the consequences.

Environment A word I hate. Supposed to mean the flavor of hardware, software and operating system you have. Misused so much I've lost all respect for it.

EOF End Of File. Special character that marks the end of a file or other document. Used in both stored and transmitted data.

EPS Encapsulated PostScript. An image description format. EPS translates graphics and text into descriptions to a printer of how to draw them. The font and pictures themselves need not be loaded into the printer; they've been "encapsulated" into the EPS code. The EPS file holds all the document data as un-rendered graphics and PostScript font mathematical data. Final rendering of the graphics file occurs in the laser printer or imagesetter, and is optimized to take advantage of the particular printer.

Erasable optical Another word for rewritable. An optical disc that does not mark its data permanently into the disc surface.

Erasable storage A storage device whose contents can be changed, i.e. random access memory, or RAM.

Erase head On a magnetic tape recorder — voice or video — this is the "head" which erases the tape by demagnetizing it immediately before a new recording is placed on the tape by the adjacent record head.

E-scale A clear plastic template the shows the letter E in descending point sizes. For judging type size.

Ethernet A local area network standard.

Exception word dictionary Typesetting programs have dictionaries that show them how to hyphenate words. You can build exceptions that overrule the standard dictionary, for words you want to hyphenate in a particular way.

Executable A program that can be run. In the PC world, a .EXE file.

Expanded characters Spread out letters, like those printed on early dot matrix printers.

Extended partition Created from free space on a hard disk and can be subpartitioned into zero or more logical drives.

Extension The period and up to three characters at the end of a filename. An extension usually indicates the type of file or directory. For example, program files have default extensions of .COM or .EXE. Many applications use a default extension when you save a file the first time. For example, Windows NT Notepad adds. TXT to all filenames unless you specify otherwise.

Extraction The string of data selected from printed matter that is used as the eye-readable label on the microfiche card.

Extrusion In graphic design, the process of creating a three-dimensional object by taking a two-dimensional shape and pushing it back through the third axis. It is often used to add depth to text, or to create a cube from a square.

Eye-readable Images recorded on a microform which can be read without magnification. This is a relative term, as it depends on the quality of the eye in question.

Moore's Imaging Dictionary

F The sixth letter of the alphabet. In ASCII, uppercase "F" is represented as hexadecimal 46; a lowercase "f" is hexadecimal 66. In EBCDIC, an uppercase "F" is hexadecimal C6; a lowercase "f" is hexadecimal 86.

Face Typestyle, as in typeface.

Facets In graphics programs, simple objects (called "primitives") are often drawn as polygons. Each surface of the object is a facet.

Facilities An enormously imprecise term that means "anything and everything." To me it sounds like toilets. But it's not. Usually means the equipment and devices you need to create your product. At this publisher's office, the facilities are computers and networks; at a steel mill the facilities are furnaces and cooling equipment. The significance is: buying and managing facilities requires a lot of paperwork and organization. Image-based facilities management software is a hot area.

Facsimile Long for fax.

Fade to black A video editing transition in which an image smoothly fades to a black screen (also called a fade-out transition). Commonly used to signify the end of a scene.

Fade-out See fade to black.

Fanfold Continuous tractor feed paper, also called "green bar paper."

FAT File Allocation Table. Data written to a magnetic disk is not necessarily placed in contiguous tracks. It's usually divided into many clusters of data in many locations on the disk surface. The FAT is the special area on a disk which keeps track of where clusters of data have been written for retrieval later.

2. In video editing, recording, logging or marking a clip that is longer than needed, i.e. adding "fat." This is done to make sure there is enough video and audio to trim the edit points to exactly the right frame.

FatBits An option in MacPaint, the popular graphics program for the Mac, that allows the user to "enlarge" a bit-mapped image and alter it pixel-by-pixel.

Fault tolerance The ability of a system to recover from an error, a failure, a disaster or loss of power. True fault tolerance means full automatic recovery without disruption of user tasks or files.

Fax Slang for facsimile. A collection of technologies, really. Facsimile first scans, then digitizes, then compresses the image of a paper document. It then converts the digital image to analog form with a modem. The fax machine then dials and arranges a data communications session — agreeing on speed of transmission and protocol — with a remote machine. The analog version of the document is then transmitted. Meanwhile, the receiving machine captures the analog data, reconverts it to digital form AND finally prints a copy, or facsimile, of the original document.

Common facsimile technology is in its third and a half generation, called Group 3 enhanced. Group 3 transmits a page at 14,400 bps (with data compression) in less than half a minute. Group 3 resolution is 203 x 98 dpi in standard mode and 203 x 196 dpi in fine mode.

Fax-back A service that lets a user call a phone number, hear voice prompts, conduct a document search and have a document sent to his fax machine. Also called fax-on-demand.

Fax board An add-on circuit board, that fits in a PC, that sends computer files in fax format to either a fax machine or another fax-board equipped PC. Quality of the image is better, since it isn't scanned. And when the output is directed to a laser printer, the image is even sharper and comes out on regular paper (as opposed to the dubious quality of fax machine's thermal paper).

Moore's Imaging Dictionary

Fax-on-demand A service that lets a user call a phone number, hear voice prompts, conduct a document search and have a document sent to his fax machine. Also called fax-on-demand.

Fax server A specialized interactive voice response system which sends facsimile messages to a fax machine you designate by touch-toning in numbers.

The fax machine you designate might be the one you're calling from, or another one.

Faxes The verb for fax. "He faxes his love letters."

Faxs More than one fax. "She gets 20 love faxs a day."

FCB Forms Control Block. Information at the beginning of a set of data to be sent to a COM recorder that contains formatting controls.

FDDI Fiber Distributed Data Interface. High bandwidth (100 Mbps) fiber network. Enough bandwidth to transmit data, images and voice together.

FDM Frequency Division Multiplexing. A technique in which the available transmission bandwidth of a circuit is divided by frequency into narrower bands, each used for a separate voice or data transmission channel. This means you can send many more transmissions on one circuit. FDM is still the most used method of multiplexing long-haul transmissions. It is typically used in analog transmission.

Feathering 1. Adding a small amount leading between lines to make a column justify horizontally. 2. Softening the edges of a graphic item that's been cut and pasted onto another one for a more realistic appearance.

Feature extraction The optical character recognition (OCR) technique used by "omnifont" OCR software. The software keeps libraries of information regarding characters' features, called "experts." These experts look at the portions of the character they're responsible for: the letter has two diagonal lines (notes one expert); the lines intersect at the top (notes another); it has a horizontal line that crosses from one of the lines to the other (notes a third). The experts' opinions are combined, and most of them agree that the features they've extracted are common to the letter "A". It must be an "A". Feature extraction CAN be used to recognize handprinting, in certain constrained cases. Handprint recognition, though, is most often the realm of "neural network"-based OCRs.

Ferric chrome A tape comprised of a layer of ferric oxide particles and a layer of chromium dioxide particles and combining the attributes of both.

Ferric oxide A tape whose coating is of red iron oxide. The original material used for magnetic recording tapes.

FGO Federal Government Operations

Fiber optic cable Cable made from thin strands of glass through which data in the form of light pulses is transmitted. Excellent, but very expensive, for very high-speed transmission over medium to long distances.

Fiche Short for microfiche.

Fiche break The place where one set of data is separated from the next on a microfiche. May contain a few blank pages, or may instruct the reader to advance to the next column, or to the next fiche.

Field 1. The smallest logically distinguished unit of data in a record, as in "There are 12 fields in that record." "Logically distinguished" means there are similar units of data in other records that have something in common. For example, "STREET" is a field, the entire address could be a record. All the address records together is a database.

2. In Windows, the field is the empty line in a dialog box where you enter data.

3. In NTSC video transmission, each frame is displayed in two interlaced signals, called fields.

Field separator The prearranged code, typically a comma, that separates fields in a record. Also called a delimiter: "The records in that database are comma-delimited."

FIFO First In, First Out. Queue handling method that operates on a first-come, first-served basis.

File All the data that describes one document or image, maintained under a single naming code and stored in a computer or in a storage medium.

File cabinet paradigm Most document management packages

strive to be as easy to use as possible. One ploy is to create an interface that mimics the way documents are usually handled manually. Therefore electronic filing is usually based on a file cabinet paradigm: there are software "cabinets," "drawers" and folders," usually represented pictorially with icons. The cabinets and such are only logical links. Chances are the image you store in the cabinet called "accounts payable" is not physically placed in memory next to all the other images in the cabinet. It is only logically linked, via software pointers and naming schemes.

File drawer An organizing concept used in document management software that uses a file cabinet paradigm, which most do. Used to generally organize a group of related "folders."

File extensions MS-DOS files can have an eight-character filename followed by a period and a three-character file extension. While most extensions are arbitrarily assigned by users or companies, some extensions are reserved for special purposes or circumstances. Many local area networks follow the MS-DOS naming conventions:

.EXE^DOS executable file

.BAT^DOS executable batch file

.DAT^ASCII text file (usually)

.COM^DOS executable command file

.ERR^Error log file

.OVL^Overlay file

.HLP^Help screens which appear by pressing F1

.SYS^Operating system file

File format The way information is structured in a file. Applications always store data files in a particular format. A format readable by one application may not be readable by another application. They are often named by the three-chacter extensions of their names. Examples of image file formats:

AIS: Xerox RGB format
AMG: Amiga
AVI: Audio Video Interleaved, Microsoft's Video for Windows file format

Moore's Imaging Dictionary

BMP: Windows Bitmap
BPX: Limena BIGPIX format
CGM: Computer Graphics Metafile, an object file for drawing
CRF: Crosfield, a high end prepress format
DCS: A five-file EPS format (CMYK files and one low-res replacement)
DXF: Data Exchange Format for CAD/CAM
EPS: Encapsulated Postscript
GIF: Compuserve's bitmap format
HEL: Hell Chromalink
HSK: Handshake, Crosfield's format for RGB and CMYK
ILS: Adobe Illustrator
IMG: GEM Paint format in Ventura, Xerox monochrome files
JPG: JPEG photo compression format
MSP: Microsoft Paint, monochrome
PCD: Kodak's Photo CD
PIC: PICT, Apple's first graphics standard
PC2: PICT2, an update of PICT
PCX: ZSOFT's PC Paintbrush format
PHS: Photoshop
PIX: Island Graphics format
PXR: Pixar format
PNT: Mac Paint
PXL: Pixel Paint
RAW: Raw image data (from scientific instruments)
RLE: Run-Length Encoding, Microsoft's compression scheme
SCI: Scitex CT, a high-end prepress format
SGI: Silicon Graphics 3-color format
SUN: Sun Raster
TCL: A Hewlett-Packard format
TGA: TARGA, TrueVision's video format
TIF: Aldus's Tagged Image File Format
TON: Tone Panel, an Optronics RGB format
WMF: Windows Metafile — a vector format
XBP: X-Bitmap
XWD: X-Window Dump

File gap A short length of blank tape used to separate files stored on linear magnetic tape.

File maintenance The job of keeping your database files up to date by adding, changing or deleting data.

File management Maintaining control of all kinds of computer files — spreadsheets, databases, images, wordprocessing documents — with a software program that forces naming structures, displays directories and handles file moves and deletions from a friendly interface. Many of these softwares monitor your actions, and when you initiate a "save," pops up to control the way you store new files, etc. Very handy.

File-oriented backup Any backup software which instructs the computer to store information in files just as they appear on the originating computer, making restoration easier and more logical.

File protection Techniques for preventing the accidental erasure of data. There are physical file-protection techniques for storage media and usually work by preventing any further recording. A 5 1/4" floppy disk is file-protected by covering the notch on the bottom of the disk cartridge with tape. An 8" disk is protected by removing the tape, uncovering the notch (exactly the opposite!). Mag tapes are file-protected by removing a plastic ring in the center of the reel.

There are also nonphysical, or "logical," file protection schemes that are part of the operating system or applications software. These prevent chosen files from being erased, or edited. See write protect.

File server Local Area Networks (LANs) were invented to allow users on the LAN to share and thereby conserve the cost of peripherals (printers, modems, scanners) and to likewise share software. The file server is the machine on the LAN where the shared software is stored. It typically is a combination computer, data management software and large hard disk drive. A file server directs all movement of files and data on a multi-user communications network, namely the LAN. It allows the user to store information, leave electronic mail messages for other users on the system and access application software on the file server — e.g. word processing, spreadsheet. A file server should also stop more than one user accessing (and potentially changing) a file at the same time — a capability called file locking.

Fill Designated areas that are flooded with a particular color or pattern. Most paint packages let you create geometric shapes in filled form. All packages also let you fill irregular closed regions. Two types of such fills exist: A seed fill floods all connected regions with the color specified by the mouse or stylus pointer; a boundary fill floods a color until the algorithm encounters a specified boundary color.

Filler A short piece of text used to stretch copy and give the impression of serious academic endeavor. This definition is filler.

Fillet The rounded concave filler that connects the serif to the rest of the character's body.

Film A transparent substrate upon which images are recorded. To view the images, light passes through the film, the resulting light being affected by the opacity and hue of the image.

Film recorder A machine that makes slides. It exposes photographic film to raster images projected from an internal monitor. The resulting film can be processed normally to create 35mm slides or larger transparencies.

Filtering A process in both analog and digital image processing used to reduce bandwidth. Filters can remove information by limiting high and low frequencies, or averaging adjacent pixels to reduce the total number.

Finder The Finder is the part of the Macintosh operating system which controls routine file management activity. Finder is the software which is running whenever no application is in the foreground. Under System 6, Finder (as opposed to "the Finder") confusingly also means the version of the Finder which can only run one application at a time, as opposed to MultiFinder.

Finishing The post-production work on a document after the printing is done: trimming, binding, packing, etc.

Firewall A LAN term. A barrier set up to contain designated LAN traffic within a specified area. Routers and other internetworking devices use their access control capabilities to build firewalls that can, for example, keep fault from propagating throughout the entire internet.

Firmware Software instructions written permanently into a chip or cartridge, usually for upgrading or adding features to an existing computer.

First line form advance A printer feeder that brings the next preprinted blank form into the precise place to fill in the first line.

First read rate The percentage of times a bar code scanner gets it right in the first pass. A measurement of quality.

Fixed beam scanner A stationary bar code scanner over which you move the bar coded item, rather than move the scanner.

Fixed disk Another name for hard disk. So-called because it is installed in a computer and not meant to be removed.

Fixed field A database field that can accept a limited maximum number of characters.

Fixed spacing All characters take up the same horizontal space, regardless of their widths. An "i" would take up the same space as an "M." Also called fixed pitch. Opposite of proportional spacing.

Flame mail Slang term for rude electronic mail. Bill Gates, Microsoft chairman, is said to be famous for the flame mail he sends to employees between midnight and 2 AM.

Flat A composite set of printing negatives, ready for platemaking.

Flat form A software form template with no data entered yet.

Flat text Raw ASCII text, i.e. no formatting commands in place.

Flatbed A scanner design in which the document is placed on the scanner's bed, either manually or by an automatic document feeder, and remains in one position during scanning. As a result, flatbed scanners provide a more stable target than other scanner designs. But they are generally slower.

Flicker Monitors flicker when the refresh rate — measured in Hz — is too low. The rule of thumb is 70 Hz or above is "rock solid."

Floating accent In typesetting, an accent mark that is placed over the letter during typesetting. Alternatively, some fonts have a "pre-accented" letters as part of the character set.

Floating selection A selected area that is conceptually floating above the image, allowing it to be manipulated without affecting the background (for example, the contents of the Clipboard).

Flop To turn over an image to its mirror image, i.e., from facing left to facing right. Not to be confused with reverse, which means convert the image from positive to negative or vice versa.

Floppy disk A portable magnetic storage medium that can be inserted in and removed from a floppy drive. Never designed as an archival medium, and should not be used as such.

Floptical A combination of optical (laser-based) track positioning and magnetic read-and-write. It's basically floppy disk technology with very precise tracking, so it can pack more data closer together. Floptical is a registered trademark of Insite Peripherals.

Flush left/right To set columns of type aligned either on the left or the right, with the opposite side remaining unaligned, or "ragged."

Flying spot scanner A scanning device that uses a point of light passed over a document to convert it to electronic signals.

FMGT Short for File Management. Term for the formatting of microforms — reduction ratio, frames per column and row, number of index frames, etc.

f/number Or f/stop. The ratio of the focal length to the diameter of the aperture (opening for light).

Focus servo The device in an optical drive that keeps the read/write beam aligned despite imperfections in the medium or bumps and shakes.

FOD Fax On Demand.

Foil Slang for an overhead transparency. The expression "he gives good foil" reflects an executive's ability to make great presentations using overhead transparencies. In the '70s and early '80s, so many managers at IBM made presentations that some senior executives actually got overhead projectors built into their desks.

Folder The basic element in a file management scheme that uses a "file cabinet" paradigm, which most do these days. A folder holds sets of directories. A folder can hold other folders. It is basically a hierarchical tree-directory scheme, just like DOS's directories and sub-directories. Most document management programs use folders as another name for directories. The Macintosh operating system's file management system is based on "folders."

Foley Adding sound effects in a studio to enhance onscreen sound. The foley artist adds footsteps, door slams, gunshots, etc., that either didn't get picked up by the location microphone, or needed some extra emphasis.

Folio Noun: the page number that appears as part of the running foot matter on each page of a publication. Verb: to set the page number on the left edge of left-hand pages and the right edge of right-hand pages.

Font All the characters and digits in the same style and size of type.

Font stress Describes whether a font is roman or italic.

Foot The bottom of the page. The top is the head.

Footprint The physical area a machine occupies on your desk; the amount of square feet, or "real estate," devoted to a machine.

Foreground processing An application that requires the full dedication of the computer's processors, and won't allow another task to be performed.

Forklift upgrades Equipment that needs to be completely replaced — no enhancement or simple fix will do.

Form 1. The prearrangement of areas on a document, either paper or electronic, that contain predictable classes of data. A form forces the user to place the date in one place, the purchase order number in another, the signature in another, etc. The significance: OCR/forms processing programs can locate these data areas, or "zones," and retrieve data for input into a database. The challenges of forms designers: helping forms-processing programs locating the pertinent data from incoming forms; isolating the blank areas so the forms processor extracts only the data, not form material itself; forcing the user to fill in the form properly.

2. The added-on border of a microform, which may consist of a simple border or may contain identifying matter such as logos or titles.

Form factor The size of a mechanism, usually refers to units meant to be installed in a PC or workstation. If a new hard drive is six inches deep, and your PC can only accept five inches, then you say its "form factor" is too large.

Form feed A ability of a printer to move continuous rolls of paper into the proper position to print one page after another.

Form slide A piece of glass that is exposed along with a document to create the form, or border. Mechanical form slides have fallen out of use in favor of electronic forms, which require less handling and are less susceptible to skewing.

Formatted data Data which has been processed with software to attach the necessary titling, indexing and job separation instructions.

Formatter A wordprocessor that has features to control the appearance of text on a printed page — alignment, bolding, font sizing. Most true wordprocessors are formatters. Contrast with text editor.

Formatting Preparation of a storage medium — defining tracks, check for bad sectors, etc.

Forms processing The ability for software to accept scanned forms and extract data from the boxes and lines to populate databases. Usually based on "zoned" OCR, and often includes the ability to "drop-out" the form itself to improve OCR accuracy and save storage space.

Forward prediction A means of compressing video by creating a single compressing frame based on the differences between two consecutive frames.

Fourier transform Converting data from the space or time domain into the frequency domain. It represents data as the sum of a series of waves — amplitude, frequency and phase relationships. Used in image processing and machine vision.

Fourth generation language 4GL. A language that commands the computer at a higher level abstraction than normal high-level languages.

Fox message The test message "The quick brown fox jumps over the lazy dog." It contains every letter of the alphabet.

Fractals A third (along with raster and vector) but quite more esoteric means of defining graphics in a computer. Instead of dots or lines, fractal graphics translate the natural curves of an object into mathematical formulas, from which the image can be later constructed.

Fractal geometry was introduced in 1977 by Benoit Mandelbrot. Its greatest promise to imaging is in image compression. Mandelbrot showed that patterns in nature (and in images) consist of infinite repetitions of basic shapes. A fractal may have great detail and size, yet appear similar, even identical, in a variety of scales. The rule used to sum and scale a fractal can be expressed mathematically, the final complex form can be characterized by one empirically derivable number. The potential compression is enormous. Its science is beyond me, having come from study of the chaos of the universe. My friend Bill Chu, editor of the New York AIIM chapter's newsletter, lent me this definition.

Fractional T-1 Any data transmission rate between 56 kbps (DSO rate) and 1.544 megabits per second (full T-1). Fractional T-1 is popular because it's typically provided by a phone company (local or long distance) for less money than a full T-1. FT-1 is typically used for LAN interconnection, videoconferencing, high-speed mainframe connection and imaging. Fractional T-1 is typically provided on four-wire (two-pair) copper circuits, as is full T-1.

Fragmentation Adding and deleting records in a file creates holes like a Swiss cheese in the file. The operating system stores the data for an individual file in many different physical locations on the disk, leaving large holes between records. Fragmented files slow system performance because it takes time to locate all parts of a file.

Frame In microforms, the area of film exposed to light, along with the form, i.e., border, logos, whatever. In computer graphics, the border of the text or picture box. In animation and film, an individual picture representing a moment in time. A series of frames displayed in sequence forms a movie.

In video, conceptually, the same as one frame of motion picture film except a video frame is recorded electronically on videotape. Each television video frame actually consists of two fields, numbered one and two (or odd or even) which are interlaced, to produce a video image. NTSC uses 30 frames (60 fields) per second to create the image and simulate motion. The European PAL system uses 25 frames (50 fields) per second. Film runs at 24 frames per second.

Frame accurate The ability of a VTR or editing system to position a frame of video accurately relative to another frame of video. Necessary for video editing. You can't use your home VCR for heavy editing work unless it is frame accurate.

Frame address Indicates the location of a video frame. Used in controlling interactive presentations and in editing. In optical CAV, the frame address is a number. In CLV, it is in elapsed time from the start of the active program (in hours/minutes/seconds).

Frame buffer A large section of memory used to store an image to be displayed on-screen as well as parts of the image that lie outside the limits of the display.

Frame grabber A board-level (usually) device that changes a video picture into a digital computer graphics format. Video editing software uses grabbed frames from live video to make icons for arranging on a storyboard.

Framing bits In data transmission, the bits that separate characters in a bit stream.

Free page A page in a document that does not conform to the layout grid elsewhere throughout the document. Plays havoc with certain zone-based OCRs.

Freeze A video editing effect which displays a single frame or picture for a period of time, thus freezing the action.

French Spacing Adding an extra space between sentences. No longer used except in typewritten copy.

Frequency The number of cycles-per-second (the usual unit of time measure) of a signal or vibrating medium. Usually expressed in Hertz (Hz). In graphics, the number of lines per inch in a halftone.

Front key compression A text compression technique that works on alphabetized text files. The idea is that the "prefix redundancy" (the first few letters that repeat) can be encoded, leaving only the unique parts.

Fugitive glue A rubbery tacky substance used by printers to affix stuff temporarily into magazines. You can pull something fugitive-glued out of a magazine without tearing anything.

Full duplex A data communications scheme that permits simultaneous transmission in both directions.

Full flush Typesetting in which both the left and right sides of a column align. To do this, the typesetter must justify — insert and delete extra space — the text. Also called "flush left and right," "justified" and "full measure."

Full reversal processing The process used to get negative-reading COM originals. The steps involved are: 1. first developer; 2.

bleach; 3. clear rinse; 4. re-expose; 5. second developer; 6. fix bath; 7. wash, and; 8. dry.

Full text search The ability to search text files for occurrences of certain words, digits, sentences or patterns of characters. Scanned documents are often OCRed to create an associated text file for the purpose of full-text searching. I use it as a verb because the alternative — searched using full-text retrieval software — is too long and awkward. Full-text search is used heavily in applications like litigation support, where data is unstructured and appears randomly.

Function A subroutine.

Fusing Using heat, pressure of chemicals in a printing device to set the toner in place on the paper.

Fuzzy logic Sets adjustable tolerance levels for how close or how loosely the test search adheres to the rules. For instance, if the search word is COWABUNGA, a "fuzzy" search could find COW****GA. Good for searching in OCRed text files that are less than 100% accurate.

G The seventh letter of the alphabet. In ASCII, uppercase "G" is represented as hexadecimal 47; a lowercase "g" is hexadecimal 67. In EBCDIC, an uppercase "G" is hexadecimal C7; a lowercase "g" is hexadecimal 87.

Gain The increase in signaling power as an audio signal is boosted by an electronic device. It's measured in decibels (dB).

Galley Text, usually in one-column format, used for proofing.

Gallium arsenide GaAs. A crystalline compound of gallium and arsenic used in making microchips. GaAs chips are faster and require less power than silicon chips.

Gamma correction Humans perceive slight changes in contrast in light areas very easily, but we see less detail in dark areas. So when scanning grayscale images, the contrast must sometimes be adjusted in a "nonlinear," i.e. nonuniform way. This is gamma correction — it increases contrast in darker areas, and lessens contrast in lighter areas of an image.

Next configure the software for your specific monitor. If your monitor is not listed, you must know its resolution and vertical refresh rate and the sync type. You also adjust for your monitor's gamma correction. This means you compensate for voltage irregularities, common in all monitors, that make some colors too intense or not intense enough. Having gamma correction in hardware means you set it only once, and it stays consistent as you change applications.

Moore's Imaging Dictionary

Gammic ferric oxide Magnetic particle used in conventional floppy disks.

Gang To group photographs or artwork together for scanning or making negatives, rather than scan or shoot each separately.

Gantt chart A bar chart used to plan and track progress and resources necessary to accomplish a project.

Garbled Corrupted data.

Gas plasma screen A display device that uses inert ionized gases sandwiched between an x-axis panel and a y-axis panel. Characterized by its orange color. Generally used for large (over 21") displays.

GDDM Graphical Data Display Manager. IBM mainframe software for graphics. It outputs to terminals, printers and plotters, and accepts scanned data.

Generation 1. Each successive copy of an original. Each successive analog video copy is one more generation removed from the original footage, and each successive copy becomes poorer in quality than the last. Multiple generation analog dubs result in reduced quality of the both video and audio. Digital video and audio recording techniques eliminate generational loss. 2. Each successive improvement in computing power and/or architecture is a generation. The newer generation is utterly incompatible with the previous one.

Genlock From "synchronization generator lock." The synchronization of a number of pieces of video equipment with a single external timing signal. In videoconferencing systems, all cameras should be genlocked together.

Gesture recognition An OCR's ability to read check marks and certain other symbols. Better known as "mark sense."

Ghosting The amount of signal left in a photoelement (such as a CCD) from the previous exposure. Also called image lag.

GIF Graphics Interface Format, pronounced "Jiff." A bitmapped image file format (such as PCX, TIFF, etc.) used first on CompuServe. Has the dubious reputation of being the format of choice for storing pornographic photographs on electronic bulletin boards. Jack Rickard, publisher of Boardwatch Magazine, wrote: "Some of the photographs are reasonably good, but most feature strikingly plain women rather artlessly photographed by those whose higher calling is more aptly found in the building trades or automotive repair."

Giga Meaning billion or thousand million. In computers, it is actually 1,024 times mega and is actually 1,073,741,824. One thousand gigas is a tera. One thousand teras is a peta, which is equal to 10 to the 15th.

Gigabits Gb. One thousand million bits. One billion bits. Or more precisely 1,073,741,824 bits.

Gigabyte GB. One thousand million bytes. One billion bytes. More precisely, 1,073,741,824 bytes. Gigabytes are considered the stratosphere of storage, but imaging applications take up huge amounts of data. It only takes 10 8 1/2" by 11" color pictures, scanned at 600 dpi, to fill a gigabyte.

Gigahertz A measurement of the frequency of a signal equivalent to one billion cycles per second, or one thousand million cycles per second.

GIGO Garbage In, Garbage Out. If the input data is wrong or inaccurate, the output data will be inaccurate or wrong. GIGO is frequently the problem with data entered by hand into computer systems. Rule of life: in a production situation, it's a hundred times to harder to change incorrectly entered data than to make sure it's correct in the first place.

GIS Geographic Information Services. Grew out of using CAD products to help draw maps. Except they aren't "maps" anymore. They're "spatially oriented information systems." And GIS is more than just a drawing of a place. It shows: geographic features are located precisely by identifiers on a coordinate grid; the relative position of features with one another, i.e., boundaries, land parcels, utility connections; and descriptive "attributes" such as who owns the land, date of last sale, age of a gas pipe that's in the ground.

These elements, when linked to a computer graphic capability for the actual drawing — makes up GIS. For the latest in GIS and "CIG" — Computer Integrated Geography — check out a great newsletter called "Maps Alive," by Bill Elliott, 303-290-8062.

GKS Graphical Kernel System. A basic graphics software for producing two-dimensional images on line graphics or raster graphics output devices, such as plotters.

Glitch The momentary interruption of electrical power. Also slang for a problem or delay. "What's the glitch?" "Who glitched this thing up?"

Global search A word processing term meaning to automatically find a character or group of characters wherever they appear in a document.

Global search and replace A word processing term meaning to automatically find a character or group of characters wherever they appear in a document and replace them with something else.

Golden mean The ratio of 3:5, commonly used in graphic design because of its pleasing appearance.

Gooey Slightly derogatory slang for GUI — Graphical User Interface.

Gothic Any squared-off, sans serif typeface. AKA blackletter.

GPI 1. General Purpose Interface. A simple communications technique for controlling video equipment. An edit controller may send a GPI "trigger" to a digital video effects device at a certain point, to start the digital effects device performing an edit. 2. Gammafax Programmers Interface. C-level programming language. Real-time applications for fax switched and gateways.

Grab To capture a video image or frame from a video or graphics source and store it onto your computer.

Gradient fill In graphics, having an area smoothly blend from one color to another, or from black to white, or vice versa. There are many variations on this theme. Most programs let you apply textures, and others have "smart" gradient fill routines that lend a three-dimensional appearance. It's sometimes possible to manipulate the "ramp" — the gradualness of the blend — of a gradient fill.

Grain The particles of silver are the smallest features that can be recorded in traditional silver-halide photography. Those particles in a print or transparency cause a visible texture, called the grain, or "graininess." The finer the grain, the less visible the texture. Fine grain material is harder to work with.

Grandfather tape The first backup of a program or a data record, saved so that you can always go back to step one if something goes wrong.

Graphic input unit Pompous for scanner.

Graphic primitives Software routines that create the most basic geometric shapes — circle, arc, rectangle, square, line — which drawing packages use as foundations for the user to elaborate on.

Graphical user interface GUI.

Graphics For the purposes of this glossary, graphics are one of the three types of data that can be created, stored, retrieved and manipulated (the other two are text and documents). Graphics are basically pictures and drawings, either created by computer or entered into the computer by scanning or photographing. See vector graphics, raster graphics and bitmap for more.

Graphics-based Representation of images that, rather than use characters, uses vector or raster graphic techniques. An example is bit-mapped type. It is text, yes, but represented by graphic means. See text based.

Graphics coprocessor A programmable chip that speeds video performance by carrying out graphics processing independently of the computer's CPU. Among the coprocessor's common abilities are drawing graphics primitives and converting vectors to bitmaps.

Graphics overlay To combine computer-generated text or graphics and place in "on top" of live video.

Grayscale The range of shades of black an image has, measured from zero for black and some other number (often 255) for white. Scanners' grayscales are determined by the number gray shades they can recognize and reproduce. A scanner that can only see a grayscale of 16 will not produce as accurate an image as one that distinguishes a gray scale of 256. See "8-bit" and "16-bit."

Greek In computer graphics, the ability to represent letters and numbers as straight lines or gray areas to speed up monitor display. Used to see the overall layout, not the individual characters. On hard copy, the use of nonsense gibberish in the absence of real text as a layout tool.

Greek prefixes To the Greek word "chronous," used to mean the process of adjusting intervals or events of two signals to get the desired relationship between them. Here are the Greek prefixes that describe different timing conditions:

asyn = not with
hetero = different
homo = the same
iso = equal
meso = middle
piesio = near
syn = together

Grid A predefined array of vertical and horizontal lines used to locate data frames in a microfiche reader.

Grid location Where a frame of data is placed on a page, identified by numbers and letters assigned to the vertical and horizontal axis of the grid. For instance, a frame might be in D07 — row D, line 7.

Grippers On printing presses the metal fingers or suction units that grab paper to guide it through accurately.

Groove A continuous channel cut or molded into the surface of a recording medium to guide the reading device.

Group A collection of screen objects logically connected so they can be moved, manipulated or deleted as though they were a single object.

Group III, IV Standard compression algorithms set by the CCITT for fax, but often used in imaging. Group III and IV take advantage of the fact that a row of pixels in an image (called a scan-line) tends to bunch up black pixels with black pixels, white areas with white. Further, a row of pixels does not usually change much from one scan-line to the next. This is especially true of documents — type on a page. These bunches of pixels are counted (called Run Length Encoding) and the groups are assigned a code that represents the number of pixels (called Huffman Coding).

Group III, Group III Enhanced and Group IV are the latest coding schemes for fax transmission over public telephone lines. The Group III one-dimensional encoding scheme eliminates redundancy only within each scan-line, left to right. It does not reduce the redundancy of data between scan lines, up and down. This way, if there's an error on the line, only one scan line of pixels will be lost. You can still read the document.

Group III two-dimensional removes some, not all, of the vertical redundancy. It compresses more, but errors on the phone line causes many scan lines of information to disappear.

Compression ratios for Group III vary from image to image. Ratios of 5:1, 10:1 and 15:1 are not unusual.

Group IV compression was developed for the Integrated Services Digital Network (ISDN). Since this is a digital transmission line with built-in error detection and correction, it is virtually an error-free transmission medium. Because of this, all redundancy is removed. Very high compression. But pray there's no line errors, or you lose everything.

Groupware Software that automates a single tasks among multiple workers.

GUI Graphical User Interface. Computer control system that allows the user to command the computer by "pointing-and-clicking," usually with a mouse, at pictures, or "icons," rather than type in commands. Opposite of CUI.

Gulp Slang for a group of bytes.

Gutter The center of a two-page spread, where the pages meet in the binding.

H The eighth letter of the alphabet. In ASCII, uppercase "H" is represented as hexadecimal 48; a lowercase "h" is hexadecimal 68. In EBCDIC, an uppercase "H" is hexadecimal C8; a lowercase "h" is hexadecimal 88.

H&J Hyphenation and Justification. The arrangement of text evenly in a column (justification), usually requiring the breaking of words at their appropriate syllable breaks (hyphenation). For this feature, desktop publishing programs include large dictionaries which instruct where hyphens can be correctly placed.

Hairline 1. The smallest width rule, about 1/2 point thick. 2. The narrowest part of a character, usually a sans serif letter. 3. A thin defect — scratch — through typeset material.

Half duplex Data communications mode which permits transmission in both directions, but only in one direction at a time.

Halftone A graphic, usually created from a photograph, in which dots are used to represent continuous tones. Larger, densely placed dots which sometimes touch represent darker tones; smaller, widely spaced dots with white areas between them represent light tones. Color halftones use varying combinations of the subtractive, or "process," colors to represent full continuous tone color images. The dots are clustered in such a way that the eye and perception centers of the brain blend them together into one perceived color.

Halftones allow continuous tone photographs to be printed by conventional ink-on-paper processes. There is at present no way to print actual continuous tones. Dye sublimation comes close by allowing the edges of the dots to "spread" into one another, simulating a continuous tone. But I'm buying that dye sub is true continuous tone printing.

Halftones are created by exposing a continuous tone through a "screen" of regularly spaced holes, creating dots. Halftone quality is expressed in the "lines per inch" of the screens. The higher the number of lines, the more dots there are to represent the continuous tone. The halftones in Imaging Magazine are 133-line; the halftones in most newspapers are 65-line.

Handheld A scanner design in which the scanning head is passed over the document, paintbrush fashion. Handheld scanners are less expensive than flatbeds or sheetfeds. They have a narrow scanning area and are subject to skew, vibration and the skill of the operator.

Handle A small square at the corners and sides of a computer graphic image that represent the spot where the mouse cursor can be placed to manipulate (reduce, move, reshape) the image.

Handling zone The part of the optical disc that can be touched by a jukebox's gripping mechanism.

Handshaking Exchange of signals at the beginning of a data communications session. During this exchange, the two systems confirm each other's specs, like parity, baud rate and speed, to ensure a proper link is set for the data transmission. As with humans, once the handshaking is through, the business of communications begins.

Handprint recognition A special discipline within OCR, unfortunately and imprecisely sometimes called ICR, for "intelligent" character recognition. The current state-of-the-art is this: there is no viable solution for recognizing handwriting (cursive, script writing.) There has been minor success with unconstrained handprinting (where you scan handprint from a blank sheet of paper, with no guides for the printer). There has been enormous success with constrained handprinting (where the letters are printed in controlled boxes, sometimes with "style" samples as guides for the printer). The key has been neural networking-based OCR, which uses a dynamic, trainable database of samples as templates for characters. I've seen the courtesy amounts (the handwritten digits on checks) recognized at the rate of about two per second. That application takes advantage of many clues: the courtesy amount is usually in the same area;

it usually has a dollar sign. It also has many challenges: there are many ways of indicating "cents" on a check; and you wouldn't believe how sloppy some people are about writing checks!

Hanging indent Typesetting the first line flush with the left margin, and all other lines indented. The opposite of "normal" style, where the first line is indented.

Hanging numerals A typestyle where the descenders of some digits drop below the baseline.

Hanging punctuation Typesetting quote marks and other punctuation marks so they extend beyond the right or left margin of the type column.

Hard copy Anything on paper. It is all well and good to have information flash by on your display, but there are times when you want to take a hard copy with you. This dictionary was written on a computer screen. Now you have a hard copy in your hands. In this case, that's a lot more useful than having it on a disk.

Hard disk A direct access storage device (DASD) that uses rotating platters of magnetic recording material, sealed inside an airtight assembly. Data is written and read from a recording head, guided by a mechanical arm. Generally, hard disks are fixed inside a PC, but there are removable-cartridge versions. Hard disks store anywhere from five to hundreds of megabytes.

Hard error An error in data communications that cannot be repaired by correction schemes.

Hard hyphen A "typed-in" hyphen. Will always appear. Contrast with soft and discretionary hyphens, which print only if the justification of the column forces a word to break at the point of the soft hyphen.

Hard RAM Carve some memory out of a computer's RAM; power it continuously and bingo you have Hard RAM, also called a Virtual Disk. Setting up a RAM disk lets you use your computer's conventional, extended or expanded memory to simulate a disk drive (or drives). The primary advantages of a RAM disk are its very fast access speed and its battery power-saving properties. It has not mechanical element to slow it down or use additional power. Hard RAM exists in memory beyond 640KB.

Moore's Imaging Dictionary

Hard sectored Sector boundaries on a magnetic disk that have been physically marked, usually with a hole punched on the medium. Hard sectored disks are not very common these days.

Hard space A specially designated space bar character, which won't allow two words to break for justification. Often used for company names that shouldn't be broken.

Hardwired Originally used to indicate a fixed, permanent hardware connection. Now expanded to include a selected option — in hardware or software — which cannot be easily changed (rewired or re-programmed) by the user.

Hash total Adding up one or more information fields in order to provide a check number for error control. The addition is not intended to have any meaning other than for checking.

HBA Host Bus Adapter. A printed circuit board that acts as an interface between the host microprocessor and the disk controller. The HBA relieves the host microprocessor of data storage and retrieval tasks, usually increasing the computer's performance time. A host bus adapter (or host adapter) and its disk subsystems make up a disk channel.

HDTV High Definition TV. A proposed standard for TV broadcast. Not agreed upon yet, most proponents recommend the doubling of current 525 lines per picture to 1,050 lines, and increasing the screen aspect ratio (width:height) from the current 12:9 to 16:9, which would create a television screen shaped more like a movie screen.

Today's typical TV set in North America contains 336,000 pixels. A high definition TV set will display at least two million pixels.

Head The device which comes in contact with or comes very close to the magnetic storage device (disk, diskette, drum, tape) and reads and/or writes to the medium. In computer devices, it performs the same function as the head on a home cassette tape recorder.

Head crash The collision of the read/write head with the surface of the recording medium. Usually causes damage and results in massive loss of data. How can this happen? Quite easily, actually. The relationship of a read/write head to a hard disk's surface has been compared to a 747 flying at top speed...10 feet above the ground.

Head gap The distance — very small — between the read/write head and the recording surface.

Head thrashing A term for rapid back and forth movements of the read/write head of a hard drive.

Head-to-head Printing sheets of paper so the head, or top, of the front of the sheet is the same edge as the head of the back of the sheet. The pages in this dictionary, and all bound documents, are printed head-to-head.

Head-to-foot Printing sheets so that the head, or top, of the front of the sheet is at the opposite edge of the head on the back. That way when you read the sheet and turn it over, top to bottom, the head of the back of the sheet is at the top. Clear?

Header Data attached to the beginning of an electronic file that describes or controls the actions that should take place to the following data.

Header sheet An instruction sheet for an optical character reader which informs it of the format to expect on the following sheets which are to be scanned.

Helical scan A magnetic tape storage method that increases media capacity by laying data out in diagonal strips. Used in video tape recorders.

Hercules graphics The Hercules standard of monochrome graphics on a monochrome PC monitor is 720 x 348 pixel resolution and 64K screen memory. This encoding was never adopted as a color standard and is now pretty well obsolete.

Hertz Measurement term for frequency. Means "cycles per second."

Heterogeneous platforms What most people have, i.e., a mix of computer types, generations and brands, bought over the years with precious little planning or management. What some people euphemistically call an "enterprise network."

Heuristic associations A text-searching technique that uses frequently appearing words and their relationships in documents in an original hit list to conduct a subsequent search. Simple example: you search for documents containing the words "Clinton" and "budget." A heuristic analysis shows that most of those documents also contain the word "deficit." So the text search software runs a second search on "deficit," resulting (hopefully) in appropriate documents that you wouldn't have found otherwise.

Hexadecimal Counting system using the base of 16 — 10 digits and six letters. In hexadecimal notation, the decimal number numbers 0 through 15 are represented by the decimal digits 0 through 9 and the alphabet "digits" A through F (A = decimal 10, B = decimal 11 and so forth).

HFDL Host Forms Description Language. A Xerox mainframe term.

HFS Hierarchical File System. In DOS, the file management system that allows directories to have subdirectories, and sub-subdirectories. In Macintoshes, files my be placed into folders, and folders to be placed within other folders.

HIAM Holographic Index Access Method.

Hidden line In computer graphics, a line segment that is obscured in the two-dimensional representation of a three-dimensional object.

Hidden line removal A technique that removes hidden lines to decrease the number of bits needed to display or store an image.

High density Floppy disks which have been manufactured for high capacity. High-density 5 1/4" floppy disks hold 1.2 megabytes; 3 1/2" floppy disks hold 1.4 megabytes.

High level format Disk or disc formatting which allocates space on the disk/c for system, directory and file allocation tables.

High resolution Basically, any image that is displayed in better quality by increasing the number of dots, or pixels, per inch than normal. Usually refers to better quality computer displays, but can describe printer quality as well. Called hi-res, for short.

"High Sierra" Specification Also called "ISO 9660" and the "Yellow Book." A standard for CD-ROM data that describes how a table of contents should be organized, but not the actual file format of the data. Has led to incompatibilities among CD-ROM vendors.

Highlight Brightest part of a photograph or halftone. Also, the ability of a display to emphasize a certain area by blinking, reverse-video or coloring the section.

Histogram A bar-like graph showing the distribution of gray tones and colors in an image.

Moore's Imaging Dictionary

Hits The documents your program finds when you run a search. "I asked for all documents containing the word "imaging," and got 125 hits."

Hollerith card A punched-hole 80-column card used for storing information for input into a computer. Remember the cards telling you "not to fold, bend, punch, spindle, etc."? They were Hollerith Cards. They have fallen out of general use.

Hologram The recording of an image on film as the result of splitting a laser beam. Holograms can store enormous amounts of data. Holographic-based imaging systems, which use microfilm as a medium, are an exciting frontier in COM.

Holographic optical tracking An optical (laser-based) tracking mechanism that can locate the read/write head in very precise position by matching the alignment of several tracks with a template in its tracking head. By precisely aligning the head with the track, you can pack more data closer together. Iomega Corporation does this.

Home computer Where you used to play games, but you don't even have time for that anymore. Also called paperweight.

Homogeneous platform The rare example of a well-managed MIS strategy (or a good salesperson). It means all the computer equipment is of the same model, brand, etc. Very unusual. If you know of one, take a picture for me.

Homographs Words that are spelled the same, but have different meanings and usually different pronunciations. Example: record (noun) and record (verb). Complicates full-text searching software.

Hookemware Free software that contains a limited number of features designed to entice the user into purchasing the more comprehensive version.

Hop count A LAN definition. The number of nodes (routers or other devices) between a source and a destination. In TCP/IP networks, hop count is recorded in a special field in the IP packet header and packets are discarded when the hop count reaches a specified maximum value.

Horizontal scan rate The frequency in Hz (hertz) at which the monitor is scanned in a horizontal direction; high horizontal scan rates produce higher resolution and less flicker.

Horizontal table In indexing, a table with entries that follow one another sequentially, i.e., entry number one is byte number one; entry two is byte two.

Host Computer in which an application or database resides or to which a user is connected. Sometimes used generically as synonym for computer.

Host table An ASCII text file where each line is an entry consisting of one numeric address and one or more names associated with that address. Host tables are used to change into numeric addresses.

Hot fix A feature of Novell's NetWare LAN (local area network) operating system in which a small portion of the hard disk's storage area is set aside as a "Hot Fix Redirection Area." This area is set up as a table to hold data that has been "redirected" there from faulty blocks in the main storage area of the disk. It's a safety feature. Faulty blocks are marked so they will not be used.

Hot key combination A combination of keys on the keyboard that are pressed down simultaneously to make the computer perform a function. For example, the Ctrl, Alt, Del hot-key combination will warm boot an MS-DOS computer.

Hot redundancy A component or system runs in parallel with an identical "twin." Should one twin fail, the other is already running and provides full service without interruption.

Hot spot When a single pixel is activated by clicking the mouse on it.

Hot standby Backup equipment kept turned on and running in case some equipment fails.

Hot swap The ability to remove and replace any drive from an array, for maintenance or upgrade, without removing the array from service.

Housekeeping Computer activities that are not part of an application. File management, defragmenting, archive migration are housekeeping chores.

HP Stands for Hewlett-Packard. The name is hyphenated; the acronym is not. Go figure.

HPFS High Performance File System. IBM's OS/2 operating system can use either the FAT file system, common to MS-DOS, and the HPFS. You can mix and match each and select one at boot time, thanks to OS/2's Dual Boot option. IBM says HPFS is much more efficient than FAT. It tries to store all files on disk contiguously and uses its own built-in cache. HPFS' most notable attribute is the long, 254-character file names and case preservation. OS/2 remembers file names as upper and lower case (though it's not case-sensitive to commands).

HPGL Hewlett-Packard Graphics Language

HRI Human Readable Information. A preposterous IBM acronym.

HSB Hue Saturation Brightness. A color model. Hue (the color); the saturation (the amount of pigment); and brightness (the amount of white included). With the HSB model, all colors can be defined by expressing their levels of hue, saturation and brightness in percentages. Same as HSL (L stands for luminance) and HSV (V stands for value).

Hue The primary wavelength color on the light spectrum, i.e. its color. See value and chroma.

Huffman coding A popular lossless compression algorithm that replaces frequently occurring data strings with shorter codes. These data strings are arranged in a "tree structure," with shorter codes at the "top" of the tree. The most frequent occurrences of data are assigned the shorter codes near the top of the tree; more unique strings get lower, i.e. longer, codes. Some implementations of Huffman include tables that predetermine what codes will be generated for a particular string. Other versions build the code table from the data stream during processing. Group III and IV (fax-standard) compression use Huffman coding.

HVQ Hierarchical Vector Quantization. A method of video compression introduced by PictureTel in 1988 which reduced the bandwidth necessary to transmit acceptable color video picture quality to 112 kbps.

Hybrid A combination of technologies, whose sum is a third "mongrel" technology. An example in imaging could be COM, where computer technology is used to create microfilm.

Hyperdensity A 1" wide magnetic tape storage medium that holds 3,200 characters per (linear) inch.

Hypermedia A multimedia word. Describes multiple connected pathways through a body of information. Hypermedia allows the user to jump easily from one topic to related or supplementary material found in various forms, such as text, graphics, audio or video. For example, suppose you received a hypermedia document about a new filing system. You could click on a document icon and read a description. You could then click an icon to see an illustration of a file structure, and then click another icon to hear and see a video explaining the file system. Hypermedia is the name for both the authoring systems that builds such an application and the playback software that delivers it.

Hypertext A text retrieval method that attempts to emulate the way your brain works, i.e., by grouping related groups of text into logical "nodes" or "chunks." The nodes can be linked together based on their relationships. Nodes are also called "chunks," "cards" or "pads."

Hyperware Software that doesn't exist, announced to test the market. Even less reliable than vaporware, which has been announced and even shown to chosen customers, but has not been shipped to any real customers yet.

Hz Abbreviation for Hertz; cycles per second. Often used with metric prefixes, as in kiloHertz (kHz).

Moore's Imaging Dictionary

I The ninth letter of the alphabet. In ASCII, uppercase "I" is represented as hexadecimal 49; a lowercase "i" is hexadecimal 69. In EBCDIC, an uppercase "I" is hexadecimal C9; a lowercase "i" is hexadecimal 89.

I Used sometimes on switches to mean "on." The "off" setting is "O."

Icon The basis of a graphical user interface, an icon is a picture or drawing that represents a device or program which is activated, usually with a mouse, to access the device or run the program.

Identifier The name of a database object (table, view, index, procedure, trigger, column, default, or rule). An identifier can be from 1 to 30 characters long.

IEEE Institute of Electrical and Electronics Engineers. A standards organization.

IMA Interactive Multimedia Association. Industry association formed in 1991 to create and maintain standards for multimedia systems.

Image The digitized representation of a picture, graphic or document. Digitizing images "democratizes" them, placing them in the same data processing category as text and numbers, and bringing their users within reach of the same benefits that text- and number-crunchers have enjoyed for many years.

Image board An imaging-dedicated processor(s). Relieves the CPU from many imaging-specific tasks — compression, decompression, display, zooming, shrinking, scale-to-gray. In fact, does them better than the CPU.

Image data The resulting file, usually in a "standard" format, when an image is digitized.

Image lag The amount of signal left in a photoelement (such as a CCD) from the previous exposure. Also called ghosting.

Image processing Think of "data processing": it refers to the manipulation of raw data to solve some problem or enlighten the user in some way not possible without the manipulation. So it is with image processing. Digitized images which have been "acquired" (scanned, captured by digital cameras) can be manipulated. The purpose may be simply to improve the image — change its size, its color, or simply to touch-up parts of it. But a more important application of image processing is to compare and analyze images for characteristics that a human eye alone couldn't perceive. This ability to perceive minute variations in color, shape and relationship has opened up applications for image processing in high-speed manufacturing quality control, criminal forensics, medicine, defense, entertainment and the graphic arts.

Image processor Device that takes input data and changes it into the proper format for an imaging device — printer, display, microform, or computer.

Image resolution The fineness or coarseness of an image as it was digitized, measured as dots-per-inch (DPI), typically from 200 to 400 DPI.

Imagesetter A high-resolution imaging device specially applied to create type and graphics. Uses either raster or vector techniques to expose photographic paper or film. Contrasted with a character setter, which creates only alphanumeric characters by exposing paper or film through a mask with the shapes of the letters engraved in it.

Imaging Recording "human-readable" images — pictures, images, motion, text, etc. — into "machine-readable" formats — microfilm, computer data, videotape, OCR output, ASCII code, etc.

That's the same definition that appeared in my "Imaging Pocket Glossary," published the fall of 1991. Many people asked whether I would update it, given that I've had a couple of years immersed in imaging, editing IMAGING MAGAZINE, learning new stuff about imaging, etc. I thought very hard about it. The conclusion is: no. I can't think of a more comprehensive definition. I could elaborate on file formats, benefits to business and breakthroughs in technology. But I couldn't define the term any more succinctly. I'll keep it.

Imaging system Collection of devices and software programs that work together to capture and recreate images. At its simplest, it has an acquisition device (scanner, camera), an image processor and an output device (printer, microfilm, computer).

Impact printer Any printing device that uses a striking mechanism to impress ink onto the paper. Includes daisy-wheel and dot matrix. I suppose it also includes typewriters.

Import To "read-in" data from one application into another. Example, to import a graphic into a DTP application; to import ASCII text into a wordprocessing file.

Imposition The act of laying out several page images on a larger sheet of paper (called a "form" or a "signature") in such a way so that after folding and trimming the pages will appear in the proper order and orientation in the document.

Imprinting Printing more onto a sheet that has already been printed. Technically, printing a letter on preprinted stationery is imprinting. But the term is reserved for adding identifying data to a slick printed piece, like adding your company's address to the back of a provided brochure.

In-betweening AKA "tweening." Traditionally, this is the process of manually creating all transitional frames in order to give the illusion of movement. It's what animation painters do. Computer-based programs can generate in-between frames automatically. See morphing.

In point In video, the frame identified as the beginning of a clip.

Incremental cursor control The user-controlled function that moves the cursor in increments dictated by the application. In character-based text editing, the increment is typically one character in the horizontal direction and one line in the vertical direction.

Incremental backup Backing up only files that have been changed since the last backup, rather than backing up everything.

Incremental searching The ability of software to search for a word based on the letters you type, almost as quickly as you type them, refining the search with every letter you add to the clue. The idea is, the search function has gotten closer to your target word as you got closer to finishing it.

Incremental spacing The ability of a printer to move characters in very small amounts either horizontally or vertically, generally for aesthetic or graphic design purposes.

Interface board or card An internal add-on hardware device that mounts on your PC's chassis and allows the computer to talk to (interface with) the outside world.

Index At its simplest, it's a descriptive set of data associated with a document for locating the document's storage location. In a more complex and demanding role, indexing can be used to consolidate documents that may not be, at first glance, related, or that may be stored in different locations, or on different media.

Indexing stored documents is the great intellectual challenge in document retrieval. Anyone can scan a piece of paper to microfilm. The hard part is devising an indexing scheme that describes every possible parameter of each document for later searches, comparisons and processing. Full-text indexing solves the problem, but has a large overhead in index space.

Index color Assigning a smaller number of colors, chosen from a palette, to display a color image. A 24-bit color image can be displayed accurately with 256 colors...if they are the RIGHT 256. Building a new palette for each image is called indexing the color.

Index of cooperation Great term. It means the length of a particular fax machine's scan line. Should be the same as the receiving machine, or the image won't be perfectly reproduced.

Inferior Typesetting a character so that it is smaller and placed below the baseline. Also called subscript. Example: M_2.

Information float The interval of time between the acquisition of information and its availability to a user.

Infrared touchscreen. Works by shooting infrared light beams vertically and horizontally above the screen's surface. Your finger breaks the light beam, thus telling the software where you are pressing.

Initial cap The first character of a paragraph, set larger for graphic design purposes. The first letter of this chapter is an initial cap. If the initial cap is set so that the body text wraps around it, it is a "drop cap."

Initialize Startup process in which a device or system is prepared — or, more accurately, automatically prepares itself — for normal operation; usually returns all parameters to their default values.

Ink absorbency The measurable rate that ink spreads into the fibers of a particular paper stock.

Ink holdout The ability of a paper to prevent absorbency, and keep the ink on its surface.

Ink jet Type of printer that sprays ink onto paper through tiny nozzles to create letters or graphics.

Input/output activity Read or write actions that your computer performs. Your computer performs a read when you type information on your keyboard or you select and choose items by using your mouse. Also, when you open a file, your computer reads the disk on which the file is located to find and open it.

Your computer performs a write whenever it stores, sends, prints, or displays information. For example, your computer performs a write when it stores information on a disk, displays information on your screen, or sends information through a modem.

Input workstation The microcomputer or terminal at which paper or microform documents are scanned and computer files are entered. This is also the place where the index is assigned to the document.

Inquiry The act of asking a database to retrieve and present information.

Insert edit Video editing in which audio, video or both are inserted into material that is already recorded. Insert editing requires that both the source and record tapes contain a continuous control track. Insert edits do not record over the record tape's control track.

Intaglio printing A form of printing where the plates are indented to hold ink and then pressed against the paper. Intaglio was traditionally used to print newspaper Sunday supplements. It's very high-quality. The formulations of the inks is also very combustible; intaglio presses were often installed in buildings with glass or breakaway roofs over them, in case they blew up!

Integrated document and image management The coordinated management, use and presentation of documentation by an organization.

Interactive Any multimedia program, usually stored on CD-ROM or laser disc, that allows the user to manipulate the course of action. A simple example is a teaching aid: The program asks a question. Depending on the student's answer, the program would play the "yes, you're correct" part of the program or the "sorry, you're wrong" part.

Intercharacter spacing Adding adjustable widths of white space between characters to justify a line of text left and right. Also called interletter spacing. Interline spacing does the same thing vertically.

Interface 1. An imprecise yet unavoidable term for the methods and instructions a human is subject to when trying to interact with a machine. DOS's lonely "C:>" prompt" is stark, abstract, not-intuitive. Windows' interface (see GUI) is friendly, warm and inviting. Both are misleading. 2. A mechanical or electrical link connecting two or more pieces of equipment together. 3. A point of demarcation between two devices where the electrical signals, connectors, timing and handshaking are defined.

Interframe coding A video compression technique that tracks the differences between frames of video. Better than intraframe coding.

Interlaced Only every other line of pixels on a TV or computer terminal screen is refreshed on each "pass" (in American television, which is interlaced, every second line is refreshed 60 times a second). Interlacing thereby saves half the signal information that non-interlaced screens use. Few monitors used in imaging applications these days are interlaced.

Interleaved The system of writing to a hard disk that places data in non-contiguous tracks because of the rapidly spinning nature of a disk drive. The operating system keeps a "log" (called a FAT table) of where each sector of data is stored for retrieval later.

Interpolation A software method for "improving" the resolution of a scanner. An algorithm adds pixels in between pixels that the scanner actually "saw." It does this by analyzing adjacent scanned pixels and making assumptions about (interpolates) what the intermediate pixel(s) might be. Interpolation algorithms are also used to increase printers' resolutions, and improve image quality on a monitor when a bitmapped image is enlarged. Beware of interpolation when reading scanner and printer manufacturers' resolution specs.

Intraframe coding. Video compression technique that compresses within each frame, but not between adjacent frames (called interframe coding).

I/O Direct access devices to your computer. Input devices are keyboards or mice; output devices are monitors or printers.

I/O addresses Locations within the input/output address space of your computer used by a device, such as a printer or modem. The address is used for communication between software and a device.

I/O bound When a computer spends much of its time waiting for peripherals like the hard disk or video display, it is said to be I/O bound. If your computer is I/O bound, going to a faster CPU might make little difference. What you need is a faster hard disk or faster video card, etc.

Ion deposition A printing technology, similar to laser printing. Charges a drum with electrons which attract the toner. Paper is then pressed against the drum, picking up the toner. Very fast, full page printing technique, but the toner trends to smudge.

Ips Inches Per Second. A measurement of speed of a tape system.

Iron oxide A coating material on recording tape and floppy disks.

Interrupt request (IRQ) Hardware lines over which devices can send signals to get the attention of the processor when the device is ready to accept or send information. Typically, each device connected to the computer uses a separate IRQ, or should use a separate IRQ. Interrupt conflicts cause devices not to run and computers to lock up. Remove one by one until you get it to work.

Inverse video Displaying characters on a screen in the opposite color, or in contrasting colors, to bring emphasis or to mark a section of text for special treatment.

IOCA Image Object Content Architecture.

IOT Imaging Output Terminal.

ipm Impressions per minute, a printing press speed measurement. Some acronyms are always used in lower case, such as "ipm" and "ppm." Others are always capitalized. No one has been able to explain this to me.

ips Inches per second. A tape drive measurement of speed. Another lower case one.

ISA Industry Standard Architecture, an unofficial designation for the bus design of the IBM PC/AT.

ISAM Indexed Sequential Access Method.

ISO International Standards Organization. Actually, because it's French, it should literally be "International Organization of Standards." But that screws up the acronym.

Isomorphic Scaling in both the vertical and horizontal dimensions.

ISV Independent software vendor. Company that creates and sells application tools or software.

Italic Right-slanted style of any typeface. Used for emphasis, or for certain journalistic style reasons — book and magazine titles, foreign words.

IVHS Intelligent Vehicle Highway Systems. The dream that one day cars can be "driven" by intelligent processors that sense each other's proximity and navigated by GIS (Geographic Information Services) signals from satellites.

Moore's Imaging Dictionary

J The 10th letter of the alphabet. In ASCII, uppercase "J" is represented as hexadecimal 4A; a lowercase "j" is hexadecimal 6A. In EBCDIC, an uppercase "J" is hexadecimal D1; a lowercase "j" is hexadecimal 91.

Jabber In local area networking technology, continuously sending random data (garbage). Describes a station whose circuitry or logic has failed that locks up the network with its incessant transmission.

Jaggies Slang for aliasing. The ragged or stairstepped appearance of diagonal lines and curves.

Jam Something that cut sheet printers do more regularly than continuous form printers. This great definition from Xplor.

JBIG Joint Bi-level Image experts Group. A proposed standard for the compression of grayscale images. Uses the same arithmetic estimation algorthms (called the QM Coder) as JPEG.

Jitter The flickering of a displayed image. Sometimes the result of interlacing or a slow Hz rate. For no-flicker monitors, 68 Hz is the minimum. But studies have shown that the mind can perceive flicker (even if the eye doesn't see it) up to 100 to 120 Hz.

Job A batch. A set of work, such as a group of related documents to be scanned, that should be done together to keep things less complicated.

Job separator Also known as batch separator. Sometimes an actual piece of coded or blank paper; sometimes software. Divides one batch of related papers in a scanner's autofeeder from the next batch.

Jog To settle pages along one edge in preparation for trimming and binding. When you bang a sheaf of papers on your desk to line them up, you are jogging them. Large binding machines are said to "jog to the head" or "to the foot." That means the gathered signatures and bound-in cards are aligned along either the top or the bottom. This matters, because the jogged edge usually gets some material trimmed off to make the edge smooth.

Joins In OCR, characters that touch, or otherwise do not have white boundaries around all their edges.

Journal drive An optical disc drive that records all database activity as as it happens.

Journal printer Special-purpose printer which provides hard-copy output for audit trail and demand printing functions. Associated with hotel/motel management.

JPEG Proposed standard for still color image compression. Devised by the Joint Photographic Experts Group, sanctioned by the International Standards Organization (ISO) and the CCITT. JPEG works this way: A color image is digitized into pixels, each with a numerical value that represents brightness and color. The picture is then broken down into blocks, each 16 pixels x 16 pixels. Then it subtracts every other pixel, reducing it in half to 8 pixels by 8 pixels. The software then computes one average value for each block.

JPEG is usually a "lossy" compression process, meaning it deletes data in order to compress an image. It's acceptable for color image compression, though, because the human eye can't see all the nuances of shade and brightness anyway. Most JPEG compressors allow adjustment of the quality (Q). Q set "high" means less loss; Q set "low" creates a smaller file but lower quality. Lossy compression methods like JPEG are not acceptable for compressing numerical data (database, for example). The JPEG spec does allow for a lossless JPEG.

JPEG++ Storm Technology's proprietary extension to JPEG. It lets you adjust the degree of compression of both the foreground of the image and the background independently. In a portrait, you could compress the face in the foreground only slightly, while you could compress the background to a much higher degree.

Jukebox A device that holds multiple optical discs and one or more disc drive, and can swap discs in and out of the drive as needed. Same as an autochanger. Also called disc libraries. Falls into the "near-line" category of storage. They are not considered DASD devices, but much work has been done to compensate for their relatively slow retrieval times.

Jukebox management In a network, tasks like retrieval and writes to a jukebox come randomly from all the users. These tasks vary in urgency — retrievals are higher priority than writes, for example. Jukebox management software sorts out requests from the network by priority.

Management also enhances the performance of a jukebox, by intelligently reordering requests. For example, if there are three requests for images on platter 1 and two from platter 2 and the another from platter 1, jukebox management means the requests from platter 1 will get handled together, then go to platter two. Sometimes it's called "elevator sorting" — responding to requests in logical order, not in the order in which they were made.

Jumper Pairs or sets of small prongs on adapters and motherboards. Jumpers allow the user to instruct the computer to select one of its available operation options. When two pins are covered with a plug, an electrical circuit is completed, When the jumper is uncovered the connection is not made. The computer interprets these electrical connections as configuration information.

Justified Type that is set with both right and left margins aligned.

Julian Date The sequential day count, reckoned from the epoch beginning January 1, 4713 BC. The Julian date on January 1, 1990, was 2,446,892. Corrupted in modern times to refer also to an annual day numbering system in which days of the year are numbered in sequenced, i.e., first day of the year is 001, the second 002, the last day of the year is 365 (366 in leap years).

Moore's Imaging Dictionary

K The 11th letter of the alphabet. In ASCII, uppercase "K" is represented as hexadecimal 4B; a lowercase "k" is hexadecimal 6B. In EBCDIC, an uppercase "K" is hexadecimal D2; a lowercase "k" is hexadecimal 92.

K Short for the prefix kilo, as in Kbytes. Though casually referred to as 1,000, a kilo is actually 1,024 units, because of the base 2 counting method of computers. When it replaces the ",000" in a number, I like to use it with no space: "48K." When it is the prefix for a word, I add the space: "48 Kbytes."

Kaleidoscope A video effect which displays a live video picture many times within itself so that you see several live pictures of decreasing size on your screen.

Kanji A Japanese character set, consisting of symbols from Japanese ideographic alphabets.

Karaoke A frivolous (but sorta fun) application of multimedia: sound (the music) plus graphics (the lyrics) play in unison while someone sings along. Probably not what God invented the computer for.

Kbyte Kilobyte. One thousand bytes. To a computer, it's actually 1,024. So 16Kbytes, or 16K, is actually 16,384 bytes; 64K is 65,536 bytes, etc.

Kbps Kilobits per second. A measure of digital bandwidth.

Kernel The part of a computer's operating system that performs basic functions such as switching between tasks.

Kernel-based window system Operating system in which the software application executes and displays in the same physical machine. Examples include personal computers and Macintoshes. The advantage is speed. The disadvantage is that applications are closely tied to the system environment and are therefore not portable. Kernel-based window systems also do not allow users/developers to use the network as a means of sharing computer resources.

Kerning The adjustment of the spacing between two letters. Typesetters sometimes kern letters so tight they touch, causing OCR errors. See tracking.

Kerr Effect When polarized light is shined onto a magnetized surface, the light is reflected back at an angle. Further, the light is reflected in different directions, depending on the polarity of the magnetism. That quirk of nature is called the Kerr Effect and it is the basis of magneto-optical (erasable) discs.

Key entry To type in data. As OCR improves, key entry is becoming more of a thing of the past.

Keyboarding A real stupid word for typing.

Keyframe A specific frame in an animated digital effects or video clip, to which a set of specific attributes (size, rotation, location, clip time code, picture name, etc.) is assigned.

Keypad The set of number keys arranged together, usually on the right-hand side of a keyboard, that allows numbers to be typed more efficiently than those arrayed across the top of a normal QWERTY keyboard.

Keystone A monitor distortion where one end of the screen — either side to side, or top to bottom — is larger than the other end.

Keyword A word, number or phrase associated with a document to aid in its retrieval from storage. Sometimes called descriptors. There are often many keys used together to fully locate a document; together they are called an index.

kHZ Kilohertz (thousand cycles-per-second).

Kilobyte 1,024 bytes. Roughly the amount of information in half a typewritten page.

Kittyhawk A trade name for a line of very small hard disks manufactured by Hewlett-Packard. About the size of a cigarette lighter, a Kittyhawk today can hold more than 40 megs of (uncompressed) data.

Kludge A hardware solution that has been improvised from various mismatched parts. A slang word meaning makeshift. A kludge can also be in software. It may not be elegant and is probably only a temporary fix. It might be spelled "kluge."

Knowbots Intelligent computer programs that automate the search and gathering of data from distributed databases. The creation of knowbots is part of a research project headed up by the Corporation for National Research Initiatives, Reston, VA.

Knowledge-based Any software implementation of artificial intelligence.

Kodiaks An affectionate term for the Kodak employees who work in the company's arctic Rochester, NY, headquarters.

KWIC Key Word In Content. A way of automatically building an index for a printed document. It takes keywords from the titles of database records as they are entered, leaving out common "noise" and words you tell it to ignore.

Moore's Imaging Dictionary

L The 12th letter of the alphabet. In ASCII, uppercase "L" is represented as hexadecimal 4C; a lowercase "l" is hexadecimal 6C. In EBCDIC, an uppercase "L" is hexadecimal D3; a lowercase "l" is hexadecimal 93.

Label The name of a hard disk, given to it immediately after high-level formatting. Or of a tape, electronically written at the beginning.

Ladder An unsightly typographical condition where several lines end in a hyphen. The hyphens look like rungs in a ladder.

Laid A paper stock with small embossed parallel lines. Considered pretty good quality.

Lamp The correct term for light bulb, which non-technical folks put into their lamps.

LAN Local Area Network. High-speed transmissions over twisted pair, coax or fiber optic cables that connect terminals, personal computers and peripherals together at distances of up to a mile. Typically consists of PCs with adapter cards, file servers, printers, gateways to departmental or corporate computers and network software to run and manage it all. LANs let you share computing resources, printers, faxs, and file storage devices.

LAN aware Applications that have file and record locking for use on a network.

LAN ignorant Applications written for single users only. These are not recommended for use on LANs (local area networks). They do not have file and record locking.

LAN intrinsic Applications written for client-server networks.

LAN manager 1. A person who manages a LAN. Duties include adding new users, installing new hardware and software, diagnosing network problems, helping users, performing backup and setting up a security system. Unlike MIS managers, LAN managers are rarely formally trained in LAN management. Sometimes they're called LAN Network Managers. 2. The multiuser network operating system co-developed by Microsoft and 3Com. LAN Manager offers a wide range of network-management and control capabilities, but takes a decidedly second place to Novell's NetWare. It has largely been superseded by Windows NT Advanced Server.

LAN network manager An IBM-developed network management tool. It is a software program that runs under OS/2 and which provides management and diagnostics tools needed to manage a Token Ring LAN. A PS/2 running LAN Network Manager collects vital statistics and special management data packets on the ring to which it is connected. When multiple rings are involved, the LAN Network Manager relies on the token ring bridges and routers to help in managing those token ring LANs that are not directly connected to the LAN Network Manager station. IBM has installed software in its bridges called the LAN Network Manager Agent. The agent software acts as the eyes and ears for the LAN Network Manager station so that the station can manage the remote rings as if it were connected directly to them. If there were no such agents, managers of networks would be blind to what's going on these LANs.

LAN server IBM's implementation of LAN manager, now largely superseded by OS/2 2.1.

LANlord The network administrator.

Land The flat, unmarked area between the marks on an optical disc.

Land and groove A physical feature of optical discs, applied during manufacture, which defines track locations. The groove is recordable, the land separates the grooves and is not recordable.

Landscape Page or monitor orientation in which the page width exceeds the page length. Also called "comic," after the shape of frames in comic strips. Contrast with portrait. Some computer screens can actually work both ways. Some even have a small mercury switch in them that determines which way the screen is standing (portrait or landscape) and will adjust their image accordingly.

Laptop computer Portable computer, usually weighing less than 15 pounds. Probably the most useful gadget to come along in years. There's a new term for a laptop computer: notebook computer. A notebook laptop weighs less than eight pounds and is roughly the shape of a 8 1/2" x 11" sheet of paper. Subnotebooks are under 4 lbs.

Laser Light Amplification by Stimulated Emission of Radiation. A very narrow beam of intense, single-wavelength light used to read and write data on an optical disc. The laser that writes to an optical disc is set at a higher power than the laser used to read the data.

Laser fax A conventional laser printer that also can be used as a fax machine when combined with an optional plug-in cartridge and used with a personal computer.

Laser disc An optical disc standard that uses Constant Linear Velocity to store data on a 12" optical platter.

Laser optical System of recording on grooveless discs using a laser-optical-tracking pickup.

Laser printer Printer that uses a beam of light to electrostatically charge a drum so that it attracts toner, which is transferred to heated paper.

Laser servo Using a laser-readable tracking technique for magnetic recording. Results in greater data capacities. This is the basis of floptical.

Latency The time between when a read/write head finds the correct track, and the time it takes for the disk to revolve, moving the correct sector in place.

Latent image A residual image on the drum of a electrostatic printer or copier, caused by incomplete removal of the static charge from the previous image.

Lathe In graphic design, it's a way to create a three-dimensional shape by rotating a two-dimensional shape around a central axis. It is useful for creating rounded objects, such as glasses and bowls.

Moore's Imaging Dictionary

Launch To run a program, usually from within another, already running program. For instance, you're in a spreadsheet and you need to see an image linked with it. By clicking on the image's icon, you "launch" an imaging program, which displays the image.

LCD Liquid Crystal Display. A display that uses liquid crystal sealed between two glass plates. The crystals are "excited" by precise electrical charges, causing them to reflect light outside according to their bias. LCD displays have certain advantages. They use little electricity and react reasonably quickly — though not nearly as quickly as a glass cathode ray tube or a gas plasma screen. They are reasonably legible. Downside: They require external light to reflect their information to the user.

In newer LCDs, called active matrix displays, the circuit board contains individual transistors for each pixel, or dot on the screen. The enables the crystals to shift quickly, resulting in a higher quality image and the ability to display full-motion video. Active matrix displays in color are hard to manufacture. Low production yields are common, though improving.

Lead screw A highly accurate screw which is turned at a constant rate to move the optical heads of some laser disc and CD recording machines to produce uniformly spaced spiral tracks.

Leader Unrecorded tape at the beginning of a roll of mag tape for threading into the tape reading machine. Also, the dots or dashes use to guide the eye across a printed page to more associated text, like in a table of contents.

Leading Pronounced "ledding," it's the space between lines of printed text. OCRs have to be adjusted for the leading of a document to read it properly.

Leap second An occasional adjustment of one second, added to, or subtracted from, Coordinated Universal Time (UTC) to bring it into approximate synchronism with UT-1, which is the time scale based on the rotation of the Earth.

LED Light Emitting Diode, an electronic light source used in some imagesetters and printers. LEDs use less power than incandescent light bulbs, but more power than LCDs (Liquid Crystal Displays).

Ledger paper A rugged quality of paper that can withstand much handling.

Legacy Existing application — i.e. a program you already use — to which you wish to add imaging. Usage: "We image-enabled our legacy database program."

Legal size Standard paper size for legal briefs — 8 1/2" x 14"

Letter quality Printer output that resembles typewritten text. Nearly all printers in general use are letter quality these days. Was a more actively used term when dot-matrix printers were still state-of-the-art.

Level The strength of an audio or video signal. Adjusting video level has the same effect as adjusting both the brightness and contrast controls of a TV set. Adjusting audio level controls the volume of the audio.

Lexical Another word for text. These days it implies tabular-set material.

Lexicographic sort Alphabetical order sort that places numbers where they would be found if they were spelled out.

Library Collection of subroutines used by programmers. Often sold with compilers or in toolkits.

LIFO Last In, First Out. A queueing or inventory scheme whereby the most recent "thing" to come in is acted on first.

Liftoff A touchscreen term. Liftoff means your finger move the cursor; when you remove (lift off) your finger from the screen, it registers in the software as a single mouse click.

Ligature Certain letter combinations that appear frequently together are combined as one character for aesthetic purposes — for example, the ligature for "fi" is fi. Causes OCR errors.

Light Technically, light is electromagnetic radiation visible to the human eye. But the term also extends to electromagnetic radiation with properties similar to visible light, including the invisible near-infrared "light" (or more technically correct, radiation) that carries signals in most fiber optic communication systems. Light usually means the electromagnetic waves with a wavelength from .000075 cm. (red) to .000038 cm. (violet).

Light pen A video terminal input device consisting of a light-sensitive stylus connected by a cable to the video terminal. The user brings it to the desired point on the screen surface and presses a button. A light pen is used to select options from a menu on the screen or to draw images by dragging the cursor around the screen on a graphics terminal.

Lightguide Another word for fiber optics.

Lightwave This term now means laser communications systems shot through the air (as opposed to glass fiber). Yes, it contradicts the meaning of lightwave, above. These comm guys are hard to pin down. Also called "free space lightwave communications." Typically, a signal is radiated directly from a light transmitter to a receiver less than a mile away. Advantages to lightwave transmission: easy to install, no digging of cables, wide bandwidth, reliable, cheap, no FCC frequency clearance approvals required and the receiving and transmitting equipment occupy little space. Disadvantages: only works for a mile or so and is subject to attenuation (fading) from fog and dust. It's perfect for between downtown buildings, where installing cables is too expensive, too cumbersome, too slow, etc.

LIM-EMS Lotus Intel Microsoft-Expanded Memory Specification. A software technique that allows MS-DOS to access memory beyond one megabyte by mapping the memory into a window in an area that MS-DOS can access. LIM-EMS is one of the greatest techniques for speeding up getting in and out of programs. AST Research was also part of the team behind EMS, though its name doesn't appear in the acronym.

Line The line tool in drawing packages draws straight lines, typically from point to point. Most packages let you continue lines in a fashion that permits rapid creation of polygons.

Line art An image with no grayscale — only black and white. A pen-and-ink drawing is line art. Text is also a kind of line art.

Line conditioner A device that protects computers and other electronic gear from electrical surges and sags.

Line for line An instruction to a typesetter (the person, remember those?) to set finished copy exactly as shown on hard copy you've provided. Same line breaks, word wraps, etc.

Line screen The resolution of a halftone, expressed in lines per inch. Usually between 53 lpi and 150 lpi.

Line speed The rate at which data is transmitted, measured in bits per second (bps).

Line segment In vector graphics, same as vector.

Linear distortion Monitor problem, caused by misalignment of the electron guns that illuminate the color screen pixels.

Linear storage Data stored on tape is recorded one bit after another, serially, until the tape is filled. To find files, you have to rewind and wind tapes to the correct spot. Some optical discs — particularly audio CDs and videodiscs — also record linearly. Opposite of random storage. Linear is slower than random storage.

Linear array A solid-state imaging camera, e.g. a CCD, made up of a single row of light-sensitive sensors. Used in scanners and fax machines.

Lining figures Digits that are typeset so their descenders rest on the baseline, not below it.

Link To create a reference — like a pointer — in a destination document to an object in a source document. When you link an object, you are inserting a visual presentation of the object into a destination document. The linked object can be edited directly from within the destination document. When the object changes in the source document, the changes appear in the destination document, and in all other versions that are linked to the same object. Basis for OLE — Object Linking and Embedding.

Linked Object A representation or placeholder for an object that is inserted into a destination document. The object still exists in the source file and, when it is changed, the linked objects is updated to reflect these changes.

Linotronic A brand of photographic imagesetters made by Allied Linotype Company that are PostScript compatible and have a resolutions up to 2400 dpi.

Lip synch In video transmission, sound (i.e., speech) must be manipulated to stay exactly in step with movement in a visual image (i.e., a talking head). This is because the video portion of the signal contains about 100 times more information than the audio portion — and thus takes about 100 times as long to process. Codecs must, therefore, provide adjustable audio delay circuitry to delay/equalize the two signals. That's lip synching.

Moore's Imaging Dictionary

Lithography Common type of printing where the image is etched into a plate, onto which ink is applied and put in contact with the paper. Slight variation on the theme is offset lithography, where the ink is applied to a rubber blanket from the plate. The rubber blanket is then pressed against the paper. Nearly all commercial printing is lithography, and much of it — especially large run documents like magazines, newspapers and books — is offset lithography.

LL Light lens. A photocopier term.

Load 1. As a noun, the amount of work that needs to be done. 2. As a verb, to write programs into memory.

Load sharing Using networked computers to divide processing tasks on one job.

Local regulation When objects on a monitor change size in proportion to their brightness.

Locality A measure of how close commonly accessed files are to one another on a disc. "High locality" means the files reside on sectors or tracks which are close to each other. When this is the case, seek times are shorter than average.

Localtalk Apple Computer's proprietary local area network for linking Macintosh computers and peripherals, especially LaserWriter printers, is called Appletalk. Localtalk is Appletalk's LAN hardware. Appletalk is a CSMA/CD network that runs at 230.4 kilobits per second and is therefore, incompatible with any other local area network. It is also a lot slower than the present top speeds of Ethernet (10 Mbps) and Token Ring (16 Mbps). Outside manufacturers, however, make gateways which will connect an Appletalk LAN to other local area and telecommunications networks.

Logical A feature that's not physically present, but applied by software. Sectors on a hard disk are physically arranged contiguously; logically, sectors may be placed anywhere on a hard disk, requiring a software program to arrange them in the correct order.

Logical page The area in a desktop publishing documents where the text and art appears. May not be the same as the physical size of the page.

Logic-seeking printer Printer that skips over blank spaces to speed printing.

Login Procedure used to gain access to the operating system. See also home directory.

Loop A magnifying glass for inspecting small features in an image, such as dot pattern, screen angles, etc.

Looping In video editing, the process of replacing the dialog recorded on-location with dialog recorded in the studio. Also called ADR — Automatic Dialog Replacement.

Loose Having lots of white space between the letters.

Lo-res Short for low resolution. Low quality reproduction because of a small number of dots or lines per inch.

Lossless Image- and data-compression applications and algorithms, such as Huffman Encoding, that reduce the number of bits needed to represent and image (compression) but can restore 100% of its original data.

Lossy Methods of image compression, such as JPEG, that reduce the size of an image by disregarding, and losing forever, some of its original data.

Lowercase The small, non-capital letters in a font.

Lpi Lines per inch; measure of resolution for halftones.

Lpm 1. Lines per minute; one of the parameters by which electronic printers and scanners are judged. 2. LAN Print Manager.

LQ Letter quality.

LRC Longitudinal Redundancy Check. An error-checking technique based on an accumulated collection of transmitted characters. An LRC character is accumulated at both the sending and receiving stations during the transmission of a block. This accumulation is called the Block Check Character (BCC) and is transmitted in the last character in the block. The transmitted BCC is compared with the accumulated BCC character at the receiving station for an equal condition. When they're equal, you know your transmission of that block has been fine.

LSI Large scale integration. A chip with as many as 20,000 transistors on it.

Luminance The brightness portion of a video signal. The luminance of a pixel determines its brightness on a scale from black to white. See also chrominance.

LUN Logical Unit Numbers. Part of the ANSI SCSI standard that allows seven subsets of each of the seven SCSI addresses. It means there can theoretically be 49 devices on attached to each SCSI host adapter. In mid 1993, ATTO Technology introduced a means of using LUNs to raise the number of SCSI devices on a chain to 49.

LZW Lempel-Ziv-Welch. A lossless data-compression algorithm.

Moore's Imaging Dictionary

M The 13th letter of the alphabet. In ASCII, uppercase "M" is represented as hexadecimal 4D; a lowercase "m" is hexadecimal 6D. In EBCDIC, an uppercase "M" is hexadecimal D4; a lowercase "m" is hexadecimal 94.

M and m Capital M is the abbreviation for the prefix mega-. Lowercase m is the abbreviation for the prefix milli-. Mega- means one million, as in megabyte. Milli- means one thousandth, as in millisecond.

Mac Slang for the Macintosh personal computer. It's predictable that a computer as "friendly" as the Macintosh would evoke such personal feelings from users. The Mac's graphical user interface (GUI), its spectacular graphics and its easy and transparent networking ability has made it a popular desktop publishing tool.

Machine language Binary (digital) instructions that can be interpreted by a computer.

Machine readable Data which is in a format, such as ASCII, or on a medium, such as disks, tapes, optical discs or punched cards, that a computer can understand. Same as computer readable.

Machine vision Combinations of hardware (camera, computer) and software (image interpretation) for visual inspection tasks that are too difficult for humans because of speed, accuracy, repeatability or safety/environmental reasons. Usually applied in quality control. For instance, the camera takes a picture of a turbine propeller, and the software can determine whether it shows any cracks or signs of stress.

Functions of a machine vision system include: location, inspection, gauging, identification, recognition, counting and motion tracking.

Macro A written set of instructions that a computer treats as though it were an executable program, doing the all the typing and instructing for you.

Magenta 1, One of the colored inks used in four-color printing. One of the subtractive process colors; reflects blue and red and absorbs green. 2. A color image compression technique developed at Los Alamos National Laboratory and patented by Paradigm Concepts, Santa Fe, NM.

Magnetic ink Ink that can be read by a magnetic scanner; used to print account numbers on bank checks.

Magnetic recording A technique of recording analog or digital signals or data on a medium of specially prepared grains of iron oxide — good old-fashioned tape recording (although floppy and hard disks use basically the same technology).

Magnetic tape Storage medium that uses a thin plastic ribbon coated with iron oxide compound to record data with electrical pulses. Mag tape is a sequential storage medium — the next bit of data is recorded after the last bit. And in order to locate a specific bit of data, you have to look through the whole tape till you find it. One old standard for data recording is nine-track mag tape — one byte (eight bits plus a parity bit) fits across the tape, width-wise.

Magnetography A Honeywell Bull printing method that uses a drum-like disk to transfer an image to paper.

Magneto-optical A high-density, erasable recording method. A laser heats a grain of a rare earth element, which makes it susceptible to magnetic influence. The write head passes over the grain while it's still susceptible. The data can then be read by another laser, whose light is not hot enough to change the grain's polarity.

Mailslot The open slot in the front of a jukebox, used for inserting disc cartridges into the jukebox. Looks like the mailslot on your front door.

Mainframe A powerful computer, almost always linked to a large set of peripheral devices (disk storage, printers, and so forth), and used in a multipurpose environment at the corporate or major divisional level. The term — main frame — derives from the racks that typically hold a large computer and its memory.

MAN Metropolitan Area Network. Usually involves high-speed transmissions over fiber optic cables that connect LANs together at distances of about 1 to 25 miles.

Manuscript Text that needs to be typeset.

Map To assign colors from the available palette to match as closely as possible an original color image.

Map builders The software that provides iconic representations of workflow and offers workflow process definition through the use of drag-and-drop icons.

Mark (1) On microfilm, the same as a blip — a small character printed or notched on microfilm for timing or counting purposes.

(2) On an optical disc, the pit, hole, bubble, light-reflective area or magnetic domain that signifies a written bit of information.

Mark geometry The size and shape of the mark made by a laser on an optical medium.

Mark (in/out) In video editing, selecting the beginning/ending frame of a video clip.

Marking "on the fly" Marking edit points while the source or record VTR is in "PLAY" mode (vs. pausing the VTR to mark the desired edit point).

Masking Protecting part of an image from change while manipulating the area around or behind it.

Mass storage Applications, such as imaging, and processing-intensive operating systems, such as Windows, pushed the demand for mass storage options — optical discs, tape drives, arrayed hard drives. Recently, the ante has been upped further with libraries and changers — multiple arrays of already-quite-large devices like tape drives. It's not uncommon for a typical office worker to be connected (or have access to in sub-second times) multiple gigabytes of storage. That's incredible. We've run 20-platter optical jukeboxes off laptop computers!

Math coprocessor A special-purpose microprocessor which assists offloads mathematical calculations from the main CPU.

Practically required for recalculating large spreadsheets. Intel included a math coprocessor with its 486DX chip, but removed it for the 486SX. No other Intel chip has a math coprocessor built in. When you buy a math coprocessor make sure it's the same speed as your existing processor.

Matrix In imaging, we mean an arrangement of "things" (dots, LEDs, CCD elements, whatevers) in a column-by-row, x-axis by y-axis layout.

Matrix printer Same as dot matrix printer. Produces text and graphics from many small dots arranged in a matrix.

Maximize In Windows, the act of increasing the size of a window to fill the entire screen.

MByte Megabyte. 1,024 kilobytes. 1,048,576 bytes.

MCA Media Control Architecture. Apple's standard for communicating between Macs and media devices — videodisc players, CD players.

MCI Media Control Interface. A standard control interface for multimedia devices and files. Using MCI, a multimedia application can control a variety of multimedia devices and files. Windows provides two MCI drivers; one controls the MIDI sequencer, and one controls sound for .WAV files.

Measure The width of a line. Expressed in points.

Mechanical Camera-ready hard copy that is to be shot and converted to a negative for printing purposes.

Media file A file containing multimedia data, such as sound or animation.

Medium The actual material something is recorded onto. Mag tape is a medium. So is the recording layer on an optical disc. So is the vinyl your old record collection is made of.

Meg Slang for megabyte.

Megabyte Approximately one million bytes. Precisely, 1,024 kilobytes, or 1,048,576 bytes.

Megaflops Millions of FLoating Point Operations Per Second. A measure of computer performance.

Megahertz MHz. A unit of frequency denoting one million hertz, or cycles.

Memory Any means of recording data, either permanently or temporarily, for later retrieval.

Memory-resident program A program that is loaded into RAM and is available even when another application is active. Also known as a terminate-and-stay-resident (TSR) program, or pop-up program.

Menu A displayed set of options for the user in an interactive system. For instance, the list of relevant documents available after a search command has been completed.

Menu-driven A program that guides you step-by-step through a series of hierarchical sets of instructions.

Merge Combine artwork, logos, text into a final electronic document, for output or to remain as an electronic file.

Metacode Describes the formatting code, like a page description language, used by Xerox centralized printers.

Metal tape Recording tape coated with iron particles and noted for its wide dynamic range and wide frequency response.

Meter The metric unit of length, equivalent to 39.37 inches. An instrument for measuring quantities of length.

MHz Megahertz, millions of clock cycles per second. Everything that happens in a computer is timed according to a clock which ticks millions of times every second. Higher MHz computers work faster than lower MHz computers, and all the components inside the computer must be able to keep up with the system's clock speed.

Mickey Unit of mouse movement typically set at 1/200th of an inch.

MICR Magnetic Ink Character Recognition. The ability, by a scanning machine, to recognize characters printed with magnetic ink. Used on checks to help banks sort them.

Micro Channel PC bus architecture introduced by IBM in its PS/2 computers. Incompatible with its original ISA architecture.

Microcomputer A relatively small single-user desktop computer, such as a PC or a Mac.

Microfiche A 4" x 6" sheet of film containing reduced images of up to 700 pages of documents in a grid pattern, usually with a title that can be read without magnification. The number of images depends on the level of magnification needed to view them full size. For example, at 42x reduction, you can get 208 images on a fiche.

Microfilm A film medium, in tape-like rolls, for recording reduced pages of documents sequentially.

Microfloppies The latest generation of floppy disks, 3 1/2" in diameter. Invented by Sony. The microfloppy is used in the Apple Macintosh and most MS-DOS laptop computers. Used in an MS-DOS machine, a 3 1/2" microfloppy diskette will format to carry 1.44 million bytes of data — equivalent to about 500 pages of double-spaced text.

Microform Microfiche and microfilm.

Micrographics The branch of science and technology concerned with the methods and techniques for recording information on, and retrieving it from, microform. Those methods include reducing and recording images by photographic means, or directly onto film by computer (computer output microform, or COM); the location and retrieval of documents through indexing and mechanical means; and the display and magnification on display screens or paper output.

Micron One thousandth of a millimeter. Or one millionth of a meter. Equals about 1/25,000 of an inch. Awfully small. Used to measure magentic recording particles and the core diameter of fiber-optic cabling.

Microprocessor A single integrated circuit, or chip, containing a central processing unit (CPU).

Microsoft Founded in 1975 by Bill Gates and Paul Allen, Microsoft is at the time this dictionary is written the largest software company in the world. It is the originator of At Work, MS-DOS, Windows, Windows NT and Windows Telephony.

Microsoft At Work A new architecture announced on June 9, 1993. It consists of a set of software building blocks that will sit in both office machines and PC products, including:

• Desktop and network-connected printers. • Digital monochrome and color copiers. • Telephones and voice messaging systems. • Fax

Moore's Imaging Dictionary

machines and PC fax products. ● Handheld systems. ● Hybrid combinations of the above.

According to Microsoft, the Microsoft At Work architecture focuses on creating digital connections among those machines to allow information to flow freely throughout the workplace. The Microsoft At Work software architecture consists of several technology components that serve as building blocks to enable these connections. Only one of the components, desktop software, will reside on PCs. The rest will be incorporated into other types of office devices (the ones above), making these products easier to use, compatible with one another and compatible with Microsoft Windows-based PCs.

MIDI Musical Instrument Digital Interface. A protocol for sending data between musical instruments, allowing a computer to act as interface and controller of instruments, synthesizers, audio, video and even stage gear like lights and curtains. A MIDI sequencer is a software program that records the "event and gestures" of a musical performance but not the actual sounds. Kind of like a player piano roll. The sequencer data instructs, or "triggers," the instrument to play its sounds. The benefits are: 1. MIDI sequencer files are much smaller — only 1% or 2% — than the equivalent digitally recorded sound.

2. The synthesizer output is first generation, therefore top quality, every time. 3. MIDI recordings are "impermanent." They record no sound, so they can be altered, edited and rearranged seamlessly.

MIDI file A file containing all the information required to play a sound by using a MIDI device.

MIDI sequencer A program that plays or records sounds stored as MIDI files. Windows provides an MCI MIDI sequencer.

MIDI setup Specifies the type of MIDI device you are using, its channel mappings and the port it is connected to.

Migration Moving data and images from on- and near-line storage (which is expensive) to off-line storage (which is slow to get to, but cheaper.) And vice versa.

Mil One one-thousandth (1/1000) of an inch; used to describe paper and tape thickness.

Millisecond One thousandth of a second. Written ms.

Minicomputer A powerful computer shared by several uses on a network.

Minimize In Windows, the act of decreasing the size of a window so you can view another one behind it.

MIPS Millions of Instructions Per Second. A measure of raw computer speed. Refers to the average number of machine language instructions performed by the CPU in one second. A typical Intel 80386-based PC is a 3 to 5 MIPS machine, whereas an IBM System 370 mainframe typically delivers between 5 and 40 MIPS.

Mirroring A fault tolerance method in which a backup data storage device maintains data identical to that on the primary device and can replace the primary if it fails. Mirroring will typically cost you a 50% performance degradation when you write to disk and 0% performance degradation when you are reading.

MIS Management Information System. Management information that is provided by computer processing. The storage and retrieval of electronic documents is strictly speaking an MIS function. Different political environments in some companies, however, may see it differently.

Misread OCR can make mistakes two ways: it can simply not recognize a character, and translate it into some symbol that means "I don't know"; or it can translate a character, but get it wrong. The latter is a misread.

Mixer An electronic device for combining the signals from various sources onto one track. An audio mixer combines all the audio onto one soundtrack. A video mixer combines multiple video layers into one composited picture.

MJPEG (Motion JPEG) A method of video compression where each frame or field is compressed using JPEG. This may take two or four times the time of MPEG, but each frame can be accessed individually and thus it is important for editing.

MO See Magneto-optical.

MODCA Mixed Object Document Content Architecture.

Modem Short for modulator-demodulator. Device that allows digital signals to be transmitted and received over analog telephone lines.

Moore's Imaging Dictionary

Module In bar code, it's the smallest-width element, either bar or space.

Moire Pronounced mor-RAY. The undesired effect caused by overlaying dot patterns (usually halftones of photographs) which are incompatible. An independent pattern emerges when the two geometrically regular patterns intersect.

Moisture content The amount of water held in paper affects the production of a high-speed laser printer or copier. A high moisture content causes jams.

Monitor The display part of your computer system. Monitors for imaging are judged by comparing the trade-offs among two or three criteria. Maximum resolution and maximum refresh rate are the two most important. Resolution is the number of pixels, vertically and horizontally, the monitor can display. Refresh rate is how quickly the screen is "repainted" by the electron gun inside. The higher the better in both cases.

But monitors are only half of your display system; the other part is the display controller. The two must be able to take advantage of each other's capabilities. For example, it is wasteful to buy a monitor capable of 1200 x 1680 resolution if your display controller can only provide 1,280 x 1,024. By the way, it seems funny to say "only" there; I can remember thinking how awesome this Super VGA monitor looked...at 800 x 600!.

Monospacing Printers with monospacing allow the same horizontal space, called the "pitch," for all letters, regardless of their actual width. Opposite of proportional spacing.

MOON Magneto Optical On Network. Term used by CANDI Technology, in Santa Clara. CANDI stands for Computer AND Imaging Technology.

MooV Pronounced "movie." An Apple-proposed standard format for storing the basic elements of multimedia (video, animation, sound and text). Incorporated into QuickTime.

MOPS Millions of operations per second. In video processing, the more MOPS the better quality. Intel's video processor can perform multiple operations per instruction, so the MOPS rating is higher than the MIPS rating.

147

Morphing Short for metamorphosing. The process of automatically transforming one shape into another over a specified range of frames (for instance, changing a square into a circle). Morphing is similar to the polymorphic blending found in vector-based drawing programs. Some programs refer to morphing as tweening, or in-betweening.

Mosaic A video effect that "blurs" a video image by copying every nth pixel n-1 times into adjacent pixels. This gives the image a blocky appearance, making it a popular effect for creating "robot vision" (that is, making an image appear as how a robot might see it.) Also used to render a part of an image unrecognizable, like the defendant's face or some naughty bits.

Moth-eye A clever optical disc medium, sold by Plasmon Data Systems, Milpitas, CA. The disc surface is textured so that the laser marks stand out more dramatically from the surface. Thus they can place marks closer together and pack more data into less space.

Mount Extend the hierarchy of directories. Accomplished over a LAN by associating the root of one computer with a directory of a mounted computer.

Mouse Hand-driven input and pointing device for personal computers.

Mouse blur Move your mouse quickly across your screen and if you're running an LCD (for example on a laptop), the mouse's pointer will blur — due to the screen's inability to change as fast as you can move the mouse. Another term for mouse blur is submarining.

MPC Multimedia PC. The Multimedia PC Council now defines a multimedia PC as having a minimum of two megabytes of memory, a 30 megabyte hard drive, a CD ROM drive, digital sound support and Microsoft's Multimedia Extensions for Windows.

MPEG An image-compression scheme for full motion video proposed by the Motion Picture Experts Group, an ISO-sanctioned group. MPEG takes advantage of the fact that full motion video is made up of many successive frames, often consisting of large areas that don't change — like blue sky background. MPEG performs "differencing," noting differences between consecutive frames. If two consecutive frames are identical, the second doesn't need to be stored. Used to play back multimedia images from CD-ROM.

MPR II Monitors should be MPR II compliant. That means they adhere to strict safety standards regarding electric and magnetic emissions.

MS-DOS The basic command system — called disk operating system, or DOS — for IBM and IBM clone personal computers. Written and marketed by Microsoft.

MSBF Mean Swaps Between Failure. A better way to judge the reliability of a jukebox or changer. Measure of the number of disc exchanges a jukebox completes before experiencing mechanical failure.

MTBF Mean Time Between Failures. A measure of equipment reliability (the higher the MTBF, the more reliable the equipment).

MTTR Mean Time To Repair. A measure of the complexity and modularity of equipment (the higher the MTTR, the more complex — or less modular — the equipment).

Multifinder A version of the Macintosh operating system that allows several applications to be open simultaneously but with only one being in active use at any time.

Multifrequency The ability of a monitor to accept and synchronize to many different frequency inputs, making the single monitor more flexible and of higher value.

Multifunction Drive An optical drive which can use both WORM and rewritable media.

Multimedia Combining multisensory media (reading, seeing video, hearing music and speech), delivered electronically, usually to the desktop, for the dissemination of information. Often enhanced by the ability to interact with the user, i.e., change output depending on instructions from the viewer. Requires enormous amounts of bandwidth and processing power.

The benefit is more powerful communication. The combination of several media often provides richer, more effective communication of information or ideas than a single medium such as traditional text-based communication can accomplish. The reason this is all possible now is: Technologies that were once analog — video, audio, telephony — are now digital. The power of multimedia is the integration of these digital technologies.

To many people, "multimedia" is a disparate collection of technologies in search of a purpose. And it's true: most of the merger of media (as above) is taking place in business communications in the moving around of compound documents. Meantime, multimedia has moved into training and in the home for education and entertainment.

Multiprocessing Multiprocessor computer systems use many CPUs (central processing units) with one or more main memories to execute one or more series of instructions. The lure of multiprocessing is that it satisfies workgroups' seemingly insatiable appetite for processing power. Multiprocessing provides additional computer power, so that systems can run multiple jobs more quickly. Shared-memory multiprocessor systems can share resources such as disks and main memory. Tasks can run on separate local processors, sharing the same memory and disks. In addition, multiprocessing enables systems to handle larger applications.

Multistrip scan What you get with a hand scanner, which is usually about 4" wide, and a page of text, which is 8 1/2" wide. Hand scanner software can "stitch" multistrip scans together to recreate the original image.

Multisync monitor A monitor that adjusts to the type of video signal it receives. MultiSync is trademarked to NEC.

Multiprocessing A type of computing characterized by systems that use more than one CPU to execute applications. Multiprocessing is not multitasking, which is the ability to have more one application running on a system at the same time. The technique is not associated with multiprocessing, nor does it require multiprocessing to take place. Multitasking typically uses a computer with one CPU (e.g. your desktop or laptop). Multiprocessing uses a computer with several CPUs, often a server.

Multitasking When a computer simultaneously runs two or more distinct processing tasks. A computer with a single CPU can only run one application at a time. But a very fast processor, and a multitasking operating system (like Windows NT or Unix), can quickly cycle the processing, giving the appearance of simultaneous operation.

Multithreading Concurrent processing of more than one message by an application program. One of OS/2's advantages over Windows is that IBM designed it as a multithreaded operating system. Each program in OS/2 can start two or more threads, which carry out various interrelated tasks with less overhead than two separate programs would require. For example, a communications program could have three threads running: one that waits for characters to be received, another that monitors the keyboard and a third that displays information. This is more efficient than running multiple tasks because it doesn't require the overhead of an operating-system context switch.

Multiuser PC A microcomputer that has several terminals attached to it, so that multiple users can simultaneously use its resources. Multiuser PCs can either slice up the time of a single microprocessor or can give each terminal-based user his own microprocessor. Multiuser PCs are an alternative to LANs and are typically used in specialized, one-application solutions, such as a doctor's office billing system.

Multiuser software An application designed for simultaneous access by two or more users on a network. It typically employs file and/or record locking. It is not associated with multiprocessing, nor does it require multiprocessing to implement.

MVS Multiple Virtual Storage. An IBM operating system.

MVS/ESA Multiple Virtual Storage/Enterprise System Architecture. Also known as MVS/SP version 3, it's an IBM operating system that uses expanded storage for processing-intensive applications such as imaging, artificial intelligence, 4GL database manipulation.

N The 14th letter of the alphabet. In ASCII, uppercase "N" is represented as hexadecimal 4E; a lowercase "n" is hexadecimal 6E. In EBCDIC, an uppercase "N" is hexadecimal D5; a lowercase "n" is hexadecimal 95.

Nanosecond One billionth (1/1,000,000,000) of a second. Written nm. In the time it takes to read this definition, nine billion nanoseconds have passed.

Nanometer One billionth of a meter. Written nm.

Native A data file before compression. "The image is 50 megs native, 172K after compression."

Nautical mile 6,076 feet. 15% longer than a normal mile, which is 5,280 feet. A measure of distance equal of one minute of arc on the Earth. An international nautical mile is equal to 1,852 meters or 6,076.11549 feet.

NBS National Bureau of Standards. A US government agency that produces Federal Information Processing Standards (FIPS) for all other agencies except the Department of Defense (DoD).

NBS/ICST National Bureau of Standards/Institute for Computer Sciences and Technology. The NBS directorate, based in Gaithersburg, MD, is concerned with developing computer and data communications.

NDS NetWare Directory Services. A new feature of Novell's NetWare 4.0. Log onto NetWare, you're part of a group. That group gives you various "directories," i.e. access to files. Those directories may be on one or many servers.

Near-line storage Optical disc storage products that are removable, durable and have randomly accessible media. Slower access times than on-line storage. Usually refers to jukeboxes, but includes multi-function, CD-ROM and rewritable magneto-optical drives.

Near-typeset Page printers that have 300 dpi or higher resolution. To compare, real typesetters have 1,200 dpi and above resolutions.

Negative Photographic print in which tone values are reversed — white for black, lights for darks, darks for lights and black for white. Generally produced on film, for the purpose of print or platemaking.

Nested workflows Sub-workflows, represented by a single icon in a map builder, to simplify the look of the graphical representation of an overall workflow.

Nesting Another name for a hierarchical file management structure, such as DOS' directories and subdirectories, and the Macintosh's folders and sub-folders.

NET/ONE The family of local area network products, bridges, gateways, network interfaces and software from Ungermann-Bass, Santa Clara, CA.

NetBIOS Network Basic Input/Output System. A layer of software originally developed by IBM and Sytek to link a network operating system with specific hardware. Originally designed as the network controller for IBM's PC Network LAN, NetBIOS has now been extended to allow programs written using the NetBIOS interface to operate on the IBM Token ring. NetBIOS has been adopted as something of an industry standard and now it's common to refer to NetBIOS-compatible LANs.

NetWare Popular LAN operating system from Novell, Orem, UT. Since it is an actual operating system, NetWare is the link between machine hardware (file servers, printers, modems, etc.) and people who want to use that hardware. NetWare is neither DOS nor OS/2 though it can be made to look and act like them. That's part (a small part) of its popularity. Because of NetWare's power, it is not easy to learn. Most of its complexity can easily be insulated from a normal user.

Network Computer networks connect all types of computers and computer-related things — terminals, printers, scanners, modems, jukeboxes, door entry sensors, temperature monitors, etc. Local Area Networks (LANs) exist within a limited geographic area — like the few hundred feet of a small office, an entire building or even a "campus," such as a university or industrial park. There are also Metropolitan Area Networks (MANs) which typically cover a city-wide area, and Wide Area Networks (WANs) which connect computers across the state or across the country.

The benefit of computer networks can be summed up with a word: sharing. To place resources within easy grasp for many people to use, rather than redundantly supply each user his own set of resources, is cheaper, more organized and more efficient.

Network printer A printer shared by multiple computers over a network. See also local printer.

Neural network OCRs OCR software which compares characters to a large number of prerecorded samples. As new variations on a character are introduced, the neural network accepts them and therefore "learns" in a way analogous to human (neural) learning. Neural network OCRs are effectively used to recognize handprinting.

NICAD Nickel/cadmium battery. Falling out of favor. NIMH is better.

Nibble Informal term for half a byte; the first four or last four bits of a byte.

NIH Not Invented Here. The tendency of organizations to reject ideas and inventions which they didn't think of.

NIMH Nickel/metal hydride battery. A new technology developed for longer battery life.

NIS Network Imaging Server.

NLQ Near Letter Quality. A printer whose output is almost as good as a typewriter.

No read A bar code term. When a pass over a bar failed to register.

Node A point of connection into a network. In multipoint networks, it means it's a unit that's polled. In LANs, it's a device on the ring. In packet switched networks, it's one of the many packet switches which form the network's backbone. In hypertext, a node is a collection of related words and terms.

Noise 1. Interference in an audio or video signal. Audio noise may appear as a hum or hissing sound. Video noise may appear as fine white specks (snow) or streaks in the picture. 2. Irrelevant specks or marks on paper documents that degrades OCR performance.

Noise word Irrelevant or inconsequential words that are ignored by a test-retrieval index. "The," "a," "so" are almost always considered noise words. AKA stop word.

Non-impact printer A printer that uses any technique — ink-jet, thermal laser — besides having keys that strike the paper.

Non-interlaced Every line of pixels on a TV or computer terminal screen is refreshed on each "pass." In contrast, American television is interlaced, meaning only every second line is refreshed, 60 times a second.

Non-linear Storage in random access, versus the contiguous liner method of tape. Access time to data stored on non-linear devices (CD-ROMs, optical discs, hard disks) is faster than linear devices (tape).

Non-positional index A full-text index that can identify the presence of a word in a document, but not its exact location in the document.

Non-printing character A character in a transmission code which performs a control function but is not reproduced when the transmission is printed.

Non-process runout To send paper through the printer at the end of a job without printing anything on it.

Non-volatile The ability to retain information without external power applied. Hard drives, floppy diskettes and ROMs are non-volatile — their contents are preserved when the computer is off.

NOS Network Operating System. Controlling software for a local area network, which may run on top of DOS, that oversees resource sharing and often provides security and administrative tools.

NTSC National Television System Committee. The group that set standards for color telecasting which is used mainly in North America, Japan and parts of South America. NTSC uses a 3.579545 MHz sub-carrier whose phase varies with the instantaneous hue of the televised color and whose amplitude varies with instantaneous saturation of the color. NTSC displays 30 frames per second, 525 lines per frame. Each frame is divided into two interlaced fields or 262 1/2 scan lines each.

NuBus The Macintosh expansion bus. Not all Mac are "modular," i.e., have a NuBus expansion interface. Often used for video cards and modems.

Numeric Text or data made up of the digits 0 — 9 only.

Numeric machine print recognition OCR that has been restricted to recognizing only numeric characters, greatly improving its accuracy on numbers-only text. How? By limiting the alternatives. OCR works by guessing; if something has the characteristics of a B, it also has many of the characteristics of an "8." By removing the possibility of "B" from its guessing, OCR can hone in more quickly on the right answer.

Nyquist's Criterion A rule of thumb for determining the amount of digital sampling necessary to reproduce an analog wave. When applied — obliquely — to reproducing halftone photographs, Nyquist's says: for each full halftone dot, two pixels are needed horizontally and two pixels vertically to fully resolve detail. So it takes four 4 pixels per halftone dot.

An 85-line halftone screen (common newspaper resolution) has 85 halftone dots per linear inch both the horizontally and vertically. Thus, for an 85-line screen you must have an output resolution of 170 pixels per inch in both the horizontal and vertical dimensions. An output resolution higher than 170 pixels per inch would be unnecessary and therefore wasteful.

Moore's Imaging Dictionary

O The 15th letter of the alphabet. In ASCII, uppercase "O" is represented as hexadecimal 4F; a lowercase "o" is hexadecimal 6F. In EBCDIC, an uppercase "O" is hexadecimal D6; a lowercase "o" is hexadecimal 96.

O Used sometimes on switches to mean "OFF." The comparable "ON" setting is "I."

OA Office Automation. My publisher Harry Newton puts it brilliantly: "Nobody knows what it means. But there are many consultants out there who will tell you what it means for an exhorbitant amount of money." I personally think it's a benign, imprecise term for data processing when it applies to self-focused (as opposed to customer-focused) white collar-type activities — accounting, communications, document management, word processing.

OA&M Operations, Administration and Maintenance.

Object Any collection of digital data that is to be linked or embedded into another applications through Windows' OLE. An object can be a scanned image, a database file or a video or sound clip, as long as they are created in an OLE-compatible application.

Object code The binary machine code a computer understands. The result of compiling.

Object manipulation A new trend, especially in photo manipulation and editing software, to maintain each successive addition to a montage as a separate layer, forever. Takes advantage of 32-bit color

157

depth, by assigning the "first" 24 bits to represent the color channels, and the "last" eight bits to represent levels of opacity. Keeping each element — bitmaps, paint, type — separate makes it easier to experiment and try "what if" layouts, with no need to "save as" many different versions along the way.

Object-oriented programming Chunks of pre-written software routines are called objects. Object-oriented programming means reusing and reordering objects to create new programs. The object in an electronic-mail program that places messages in alphabetical order can also be used to alphabetize invoices. Programs can be built from prefabricated, pretested building blocks in a fraction of the time it takes to build them from scratch. Programs can be upgraded by simply adding new objects.

Oblique Characters which are slanted to simulate italics, but are not actually a font of italic characters.

OCR Optical Character Recognition or Reader. The ability of software to recognize and translate bitmapped scans or faxes of printed alphanumeric characters into machine-readable (ASCII or formatted) text.

Most OCRs work by using either Pattern Matching or Feature Extraction. Neural network-based (see) OCRs are just emerging. With "pattern matching," the software is given a "template" of possible characters. When the scanner sees a letter, it compares it to its library of pattern templates. If there is enough of a match, it safely assumes it has "recognized" the letter and sends the ASCII equivalent of the letter to the output file. By today's standards, pattern matching is primitive; but if you can control the font and size of incoming type, pattern matching is potentially the most accurate OCR method. Not recommended for degraded documents, such as faded copies or faxs.

"Feature extraction" is more sophisticated. Its "library" consists of groups of information (called "experts") regarding a character's features; i.e., the letter "A" has two diagonal lines; the lines intersect at the top; it has a horizontal line that crosses from one of the lines to the other, etc. The OCR compares features of the character to its experts. Feature extraction doesn't care what font the character is in, or what size it is, or whether it's bold, italic or underlined. In fact, feature extraction can recognize handprinting, if the characters are

Moore's Imaging Dictionary

carefully controlled, or "constrained."

OCR software further supports its "guesses" by knowing a little something about the language. A digit "1" is not likely to fall in between a group of letters; the letter "h" frequently follows the letter "t," etc. This is called context checking. OCR accuracy is also improved by: image prep — raising the contrast so that "noise" and "artifacts" (unwanted marks) disappear; form and line removal, which deletes superfluous rules and boxes, leaving just the clean type; and limiting "character sets," which means decreasing the amount of "possibilities" by telling the software there will only be digits, or only alpha characters, etc.

OCR's promise is its broad application. It's much more than a simple "automatic typist." OCR can enable workflow systems, by guiding a document to the correct person based on its content. It enables content-based retrieval of documents' text and images. It is an automatic data entry tool. It is the basis of extraordinary fax-based information disseminating applications.

Note: This is not future stuff. All those applications are working now. We've used them ourselves.

A note on pronunciation: We say the letters "O-C-R". We even use it as a verb: "He O-C-R-ed the page." But I just got off the phone with a guy who insisted on pronouncing it as though it were the word "OAKER." It was weird. Maybe he's right. I don't know. This is the beauty of imaging terms — usage determines how they enter the language.

Octal A numbering system with the base eight.

Octet An eight-bit byte.

Octothorp The tic-tac-toe, or "pound," button on the phone's touch-tone pad. Some people believe it's properly spelled "octathorp."

ODA Office Document Architecture. ISO's standard 8613-1/8 for document architecture and interchange format adopted by MAP/TOP 3.0, GOSIP, and standardized by ECMA as ECMA-101.

Odd parity One of many methods for detecting errors in transmitted data. An extra bit is added to each character sent. That bit is given a value of 0 ("zero") or 1 ("one") such that the total number of ones in the character (including the parity bit) will be odd.

O/E Optic to Electric conversion.

Moore's Imaging Dictionary

OEM Original Equipment Manufacturer. The maker of equipment marketed by another vendor, usually under the name of the reseller. The OEM might only manufacture certain components, or complete computers, which are then often configured with software and/or other hardware, by the reseller.

Off-line 1. Something not presently active or available for access in a system. 2. In video editing, creating an Edit Decision List using relatively inexpensive video equipment, for output on higher quality gear.

Off-line storage Archival storage not directly accessible to your computer. Requires manually finding and inserting the medium (usually tape) into a drive. Upside: inexpensive, removable, high-capacity media. Generally refers to quarter-inch cartridges (QIC), digital audio tape (DAT), or helical scan backup (8mm, 4mm and videocassette), but actually means any media stored on a shelf. Downside: s-l-o-w. And requires careful file management. See DASD, on-line and near-line storage.

Off-the-shelf Software you can buy and install today, and it will work. Sometimes called "shrinkwrapped," referring to the plastic wrap around the box. It means you don't are not meant to need intervention by a programmer or integrator to make the software work.

Office network A network within an office. An older term for a Local Area Network. User concern is with application sharing, file/database sharing, electronic mail, word processing and circuit switching.

Offset printing A form of lithography that uses a rubber blanket to accept ink from a plate, and press it onto paper. Benefits: it's fast and accurate — the rubber doesn't "slip and slide." Used for large-volume jobs.

OGL Overlay Generation Language. An IBM software that creates electronic forms for printing raw data in understandable form.

OHD Optical Hard Drive. A term pioneered by Pinnacle Micro, Irvine, CA. OHD technology, according to Pinnacle, combines the advantages of magneto-optical technology with speeds approaching those of hard drives.

OLE Object Linking and Embedding. Built into Windows 3.1, OLE lets large data files of any kind (images, sound, video clips) be either linked to, or embedded in, any Windows application that supports OLE, creating a compound document. Meaning, for example, a scanned bitmap of a person's photograph can be associated with an OLE-savvy database record or a word processing document by either: 1. Inserting a "pointer," or indicator that remembers where in storage the image is. It's not actually part of the database record; it's merely "linked." Or 2. Actually placing — "embedding" — the large file into your application file.

There are benefits to both approaches: With linking, many users can include the same image in their applications without having multiple copies. If the image is updated, the change is reflected in all the users' versions automatically.

By embedding the image, you make a copy of the image as it appears now and place the copy directly into your file. It is now protected from alteration by other users.

Omnifont The ability of an OCR to recognize any typeface font without having to "learn" (make a template in advance) that typeface. Omnifont character recognition is another word for feature extraction (see OCR).

OMR Optical Mark Recognition. Refers to machine recognition of filled-in "bubbles" on reader service bingo cards, standardized tests, etc. Also called "mark sense."

On-demand printing Creating printed material as you need it, rather than printing in advance and storing it in a warehouse. Overcomes the cost of storage, and keeps material up-to-date.

On-line 1. Something active or available for immediate access in a system. 2. In video, editing for final production with top-quality video decks and controllers.

On-line storage Randomly accessible storage (usually magnetic) with the fastest access times. Includes: hard disks, network disk subsystems, removable hard disk drives and RAID (redundant array of inexpensive disks) systems. See off-line and near-line storage.

One-dimensional coding The data compression technique, used in Group III fax machines, that treats each horizontal scan line individually, without reference to the previous scan line. The idea is that a transmission error will cause only one lost scan line. Not enough to affect readability.

OOP Object Oriented Programming. Object oriented programming is a form of software development that models the real world through representation of "objects" or modules that contain data as well as instructions that work upon that data. These objects are the encapsulation of the attributes, relationships, and methods of software-identifiable program components.

Opacity The degree of translucency of a pixel. In 32 bit color, the "last" 8 bits — the so-called "alpha channel" — describe an image's opacity. Opacity is usually measured on a scale from 0 to 100: an opacity value of 0 means an image is completely transparent and anything beneath it will show through. An opacity of 100 means the image completely covers any underlying image, with no show-through. Used in new object-based photo manipulation software.

Optical centering The human eye finds it more pleasing when something is placed slightly above the actual dead center of a page. I don't know why.

Optical disc A direct access storage device that is written and read by laser light. Certain optical discs are considered Write Once Read Many, or WORM, because data is permanently engraved in the disc's surface either by gouging pits (ablation); or by causing the nonimage area to bubble, reflecting light away from the reading head. Storage costs of optical are pretty low — around 17 cents per megabyte.

Erasable or rewritable optical discs generally use one of two technologies: "magneto-optical", which magnetically alters the bias of grains of material after they have been heated by a laser; or "phase-change," which changes the crystalline structure of a spot on the disc to reflect light in different ways.

Compact discs (CDs), CD-ROMs and laser (or video) discs are optical discs. Their storage capacities are far greater than magnetic media, and are likely to replace magnetic hard disks and tape in the near future. Their acceptance is hindered mainly by the relative slowness of their drives.

Optical disc storage and retrieval The combination of an optical drive or jukebox and software to manage the search and retrieval.

Optical scanner Same as scanner. Input device that translates human-readable or microform images to bitmapped, or rastered, machine-readable data.

Optical storage The means of storing or archiving data on optical discs such as optical WORM, CDs or laser discs. Considered near-line storage, since it's not as fast as direct access on-line storage, but not as slow as off-line storage.

Orientation The relative direction of a display or printed page, either horizontal (called "landscape" orientation) or vertical (called "portrait" orientation).

Orphan When the first line of a paragraph falls on the last line of the column. Considered unattractive typography. Typesetters pride themselves on ridding the world of orphans and "widows," small words or partial words that fall on the last line of a paragraph.

Orphaned member In RAID, a disk in a mirror or stripe set whose parity has failed in a severe manner, i.e., its head crashed. When this happens, the driver directs all new reads and writes to the remaining drives. It ignores the bad drive, "orphaning" it.

OS/2 Operating System/2. An operating system originally developed by IBM and Microsoft for use with Intel's microprocessors and for use with IBM Personal System/2 personal computers. Now OS/2 is the prime responsibility of IBM and it will run on many PCs, including those using the Intel family of PC microprocessors. OS/2 is a multitasking operating system. This means many programs can run at the same time.

OS/2 2.0 A 32-bit version of the IBM's OS/2 operating system. Apple Macintosh's operating system is also 32-bit.

Oscilloscope Electronic testing device that can display representations of analog wave forms on a cathode ray tube. A basic fixture in sci-fi movies.

OSF Open Systems Foundation.

OSI Open System Interconnect.

OSS Operating System Software.

Out-of-paper reception The ability to receive a facsimile transmission into memory when the facsimile machine is out of paper. The facsimile paper will be printed when you put in new paper. Beware: there is a limit to the size of the receiving buffer. If it fills, the fax machine will probably simply dump old faxes out to make room. No warning.

Out point In video, the frame identified as the ending of a clip.

Outline font A printer font that is described mathematically, rather than as a bitmap of dots. Outline fonts — such as PostScript and TrueType — can be scaled to any size without damaging the resolution. Once the size and shape are determined, outline fonts then have to be filled in. This is one reason PostScript is so processing intensive and can be so s-l-o-w.

Output device Any device by which a computer transforms its information to the "outside world." In general, you can think of an output device as a machine that translates machine-readable data into human-readable information. Examples: printers, microform devices, video screens.

Output file formats. Digitized images must be translated into some predictable format. Otherwise, there would be no way for a program to recreate the data back to the way it originally looked.

There are many file formats — some are common enough to be considered "standard" (don't trust that word implicitly); some are proprietary. Your computer's software must be able to accept ("support") the format, or it cannot display or manipulate the image. They are normally referred to by the extension (last three digits after the period in the file name), although they sometimes have actual names.

These are some of the raster (bitmap), vector and printer formats I've come across. I am sure there are more:

ASCII	CON	DXBTR	GEM
ATL	CRF	DXF	GIF
BMP	CUT	DXFTR	GR4AB
CALS	CVT	EPS	GROUP III
CGM	DBX	GCA	GROUP IV
CITIN	DRW	GED	HRF

Moore's Imaging Dictionary

IFF	MAC	PLT (HPGL)	TIFF
IG4	MSP	RIC	TIFF1
IGF	P10	RIC2	TIFF3
IGS	PCI	RLEIN	TIFF4
IMG	PCL	RLE	TIFUN
IMGCM	PCT	RNL	TIFWW
IMGCC	PCX	RST	VER
IMGGEM	PIC	SBP	VIFAB
JDL	PICT	SCM	WMF
KFX	PICT2	SCN	WPG
LFX	PIX	TGA	

Output hopper Place at the end of a printer where the finished documents collect. Also called a stacker.

Outsourcing A company contracts one of its internal functions to an outside company. Those functions might include handling its backfile conversion of paper to scanned images, or running the computer system. A company might be motivated to do this because they lack the internal resources (typically people). This has appeal to senior management, who are trying to reduce their uncertainties. This usually has no appeal to lower level management who might be fired.

Overcoat A transparent protective layer for optical recording media intended to protect the recorded surface from dust and scratches.

Overhead 1. A scanner design in which the original document lies flat, face up, and an optical device such as a video camera points down at it. Downside: cumbersome on the desktop; subject to ambient light and expensive. Upside: allows scanning of three-dimensional objects and books. Especially good for archiving fragile antique volumes that can't be manhandled.

2. In data communications, some data is transmitted that is not part

of the actual text, but has to do with control, addressing and error checking bits. This extra data is overhead.

Overhead bit A bit other than one containing information. It may be an error-checking bit or a framing bit.

Overhead transparency An acetate positive transparency which can be projected through a special device onto a screen or wall. Used in presentations. It is said that in the late '70s, early '80s, IBM executives did so many presentations that overhead projectors were often built into their desks.

Overlay An electronic form that displays raw data in an organized way so that humans — who are infinitely smarter than computers — can absorb information.

Overscan The image fills the screen from bezel (frame) to bezel, with no border area where the pixels aren't illuminated. AKA full scan.

Overset Text which does not fit in a specified page or text box. Overset is the scourge of publishers and typesetters. There is nothing more humiliating than to have intelligent thoughtful copy appear in print that stops in the middle of a

P The 16th letter of the alphabet. In ASCII, uppercase "P" is represented as hexadecimal 50; a lowercase "p" is hexadecimal 70. In EBCDIC, an uppercase "P" is hexadecimal D7; a lowercase "p" is hexadecimal 97.

Px64 A compression technique used in real-time videoconferencing.

Pack Another word for compressing data. Also a process used by many database programs to remove records marked for deletion.

Package What Microsoft calls the icon in an OLE image-enabled application. Clicking on the "package" launches the application used to create the object and either plays it (for example, a sound file) or opens and displays it (for example, a scanned image). No one else we've seen calls it a package; it's just the icon.

Packet A group of bits, packaged together, for transmission purposes. Three principal elements are included in the packet: 1. Control information — destination, origin, length of packet, etc.; 2. the data to be transmitted; and 3. error detection and correction bits. Sending data in packets rather than continuous streams offers more efficient use of transmission lines.

Packet switching Sending data which has been subdivided into individual packets of data, each having a unique identification and carrying its destination address. This way each packet can go by a different route. It may also arrive in a different order than it was shipped. The packet ID lets the data be reassembled in proper sequence. Packet switching is a very efficient method of moving digital data around. It is not useful for voice, yet, though experiments are underway.

Page One side of one sheet of one document. Refers to paper of any size and electronic files of any size. Can be a text page, a desktop-published page, an e-mail page, a spreadsheet page.

Page composition The act of laying out a page, placing photos and art in place, determining where text and headlines go. Layout.

Page decomposition The ability of OCR software to separate text from graphics, and maintain column/tabular formatting.

Page definition A software object containing formatting information for printing raw data in a readable way.

Page description language A means of transmitting bitmapped images from a PC application to any compatible printer. They save processing time, by sending only "instructions" to a printer, rather than the entire bitmapped image. They also allow the printer to print any font, any size. PDL is the generic term. Hewlett-Packard has been the major proponent of PDLs. They include one called PCL — Printer Command Language — with all their printers. Also called PDLs.

Page end A non-printing command character that divides pages in an electronic document into pages on a printer.

PAGEDEF IBM formatting command object for page layout.

PageMaker A desktop publishing/page makeup software program by Aldus Corporation. Runs on DOS machines and Macs. Was the first DTP product, and change the way we create paper documents. Remains, along with Quark Xpress, one of the most popular DTP programs.

Page makeup Same as layout and page composition.

Page makeup software Desktop publishing software.

Pages per minute Measurement of a scanner's speed. See PPM.

Page printer A printer that uses sheets of paper rather than continuous tractor-fed paper.

Page proof There are several ways to see a proof of a desktop published document: 1. from a desktop black and white laser printer; 2. from a desktop color printer; 3. from a high-resolution, high-quality color printer from an output service bureau; 4. from contact prints made from the finished output negs. These are in order of cost, speed and quality.

Page recognition OCR software that can tell the difference between text on a page and other items, such as pictures, artwork, etc.

Pagination Assigning the order of pages in a document, or breaking large text files into manageable pages. Not the same as imposition, which has to do with placing several ganged pages onto a single sheet in the proper order and orientation for folding and finishing.

Paint program A graphics software that simulates the act of adding colors (and often lines and type) to a blank frame. Based on pixel-by-pixel bitmap manipulation, rather than vector graphics. Different from drawing programs, which emphasize precision and automatic creation of graphs and such over the more "artistic" strengths of paint programs.

Page scrolling The ability to move from one electronic page to the next. The faster the better. Fast page scrolling often requires co-processing from a display controller card.

PAL Phase Alternate Line. The color television broadcasting system in most of Europe except France (which uses SECAM). PAL has higher vertical resolution than the US's NTSC: 625 lines per frame, 50 frames per second (100 fields per second), and a 4.43361 subcarrier. PAL and NTSC are not compatible. PAL is sometimes called "625/50."

Used mainly in Europe, China, Malaysia, Australia, New Zealand, Latin America, the Middle East and parts of Africa. PAL-M is Brazilian color TV system.

Palette The set of colors the best meet your needs when you are limited by the bit depth of the computer or application. For example, 8-bit depth can display only 256 colors. If your image is of the ocean and the sky, your palette can be filled with 256 shades of blue, thereby improving the accuracy of the result.

Pan To view a different part of a page that has been overscanned (is off the borders of the screen.)

Moore's Imaging Dictionary

PANS Pretty Amazing New Stuff. Something I say every day.

Paneling When an image is too large for the printer, you print it in panels, and paste them together. There is software that figures out where to divide an image, how much overlap to print, and labels the panels so you know how to reassemble them.

Pantone Matching System A means of describing colors by assigning them numbers. It's a de facto standard color description language adopted by commercial printers.

Paper jam What happens when you're in a hurry. Caused by skewed pages or multiple feeds. Many factors, including size, weight, thickness, even moisture content of the stock, can cause jams.

Paper memory The amount of curl a piece of paper will retain after it's been uncurled. Try this test: take a sheet of copier bond, and a page torn out of a magazine. Roll them up like a scroll for the same amount of time. Then hold them by the top edge and let them unroll. Their memories will be different.

Paper path Some printers and scanners pride themselves on a "straight-through paper path." This actually cuts down on jams. Straight through printers and scanners take up more desk space, though.

Paper sizes

US	Europe and Japan
A = 8 1/2" x 11"	A3 = 11.7" x 16.5"
B = 11" by 17"	A4 = 8.3" x 11.7"
C = 18" by 24"	A5 = 5.8" x 8.3"
D = 24" by 36"	B4 = 10.1" x 14.3"
E = 34" by 44"	B5 = 7.2" x 10.1"
	B6 = 5.1" x 7.2"

Paper tape A long thin paper roll on which data is stored in the form of punched holes. Usually used as input to other systems. Many old-fashioned telex machines still use paper tape as their storage medium. The primary benefit of paper tape is that you save on transmission line cost. The paper tape will run through at the maximum speed of the line, while a human operator typing manually would be slower. The disadvantage of paper tape is that you can't change the message once you've typed it.

Paper tape punch A device to physically punch holes in a roll of paper tape in order to store information.

Paper tape reader A device which translates the holes in coded perforated tape into electrical signals suitable for further handling. The reader may be attached to a keyboard-printer or it may be a free-standing device.

Paradigm Kind of a poncy word. It can mean using a familiar analogy to describe a more arcane subject. The best example in imaging is when marketers tell customers their document management system is "based on a filing cabinet paradigm." Also used to describe when things fundamentally change: called a "paradigm shift." According to the Economist Magazine, Thomas Kuhn invented the notion of the paradigm shift to explain what happens in scientific revolutions. A revolution happens, his theory goes, not because of startling new facts, but because of a change in the overall way the universe is perceived. After this shift, old knowledge suddenly takes on new meaning.

Parallel The transmission of bits over multiple wires at one time. Accomplished by devoting a wire for each bit of a byte. Parallel data transmission is very fast, but usually happens only over short distances (typically under 500 feet) because of the need for huge amounts of cable. Most often used in computer-to-printer, and scanner-to-computer applications. Contrast with serial.

Parallel port An output receptacle often located on the rear of a computer. Unlike serial, there is no EIA standard for parallel transmission, but most equipment adheres to a quasi-standard called the Centronics Parallel Standard.

Parallel processing 1. A computer technology in which several or even hundreds of low-cost microprocessors are linked and able to work on different parts of a problem simultaneously. 2. A computer performs two or more tasks simultaneously. This contrasts with multi-tasking in which the computer works fast and gives the impression of performing several tasks at once.

Parallel routing A workflow term. In complex processes, several steps can be performed simultaneously, and the results converge at some point afterwards. That point is called the rendezvous. Parallel processes share access to needed information.

Parity Used in error correction. A separate bit — the parity bit — is added and manipulated so that the number of 1s is odd (for odd parity) or even (for even parity). If the number of bits received don't conform to the parity, the software detects an error. Also called parity checking.

Parity bit A binary bit appended to an array of bits to make the sum of all the bits always odd or always even. See parity.

Partition A portion of a physical disk that functions as though it were a physically separate unit.

Password A string of characters you enter during a login or booting process that permits access to your level of security on a network or messaging system or bulleting board. Tips for passwords: Use all the available characters; use an uncommon family name; a misspelled or nonsense word; two words with a non-letter symbol in between. Nonsense words that alternate vowels and consonants are easier to remember because you can pronounce them. Examples: RITABILO, DISPOFEC, ANTIPILO.

Paste-up What graphic artists did in the old days. Type, graphics etc., were stuck onto mechanicals with (in most cases) a tacky wax. Rubber cement is strictly for amateurs. A few years ago at this publisher, Telecom Library, we had a room full of layout tables and a hot wax machine cooking all the time. We kept the Xacto knife people in business. Then we went desktop.

Patch A short piece of software code that overcomes some bug in the program.

Path A path (or pathname) provides directions to a file within your DOS system. It usually consists of a drive letter and one or more subdirectories, separated by back-slashes. For example, the path to this file in my computer is C:\WORK\DICTION\P. "C:" is the root directory. "WORK" is a subdirectory. "DICTION" is a sub-subdirectory to WORK. "P" is a file in DICTION.

Pattern recognition An OCR technique. Use libraries of information about how characters are built — artificial intelligence experts — in all fonts and sizes. Also called feature extraction, because the experts examine certain parts, or features, of the characters and cumulatively agree on whether the character meets enough criteria to identify it as one character or another.

PC Short for IBM Personal Computer. Used rather imprecisely to indicate an IBM or compatible, but sometimes used to indicate any personal computer. I think "DOS machine" is more specific. Also: Peg Count, Printed Circuit, Product Committee, Photoconductor, Politically Correct.

PCL Hewlett-Packard's proprietary page description language (PDL) for its laser printers. Stands for Printer Control Language. PCL Level 5, resident in the HP LaserJet Series III of printers, supports fully scalable typefaces and rotation of text.

PCM Pulse Code Modulation. The most common method of encoding an analog signal into a digital bit stream. PCM refers to a technique of digitization, not a universally accepted standard of digitization.

PCMCIA The Personal Computer Memory Card International Association (PCMCIA), a standards body made up of manufacturers of semiconductors, connectors, peripherals and systems, as well as BIOS and software developers and related industries, has developed standards for the personal computer cards (PC Cards). A PC Card is a small adapter for your personal computer, personal communicator or other electronic device. PC Cards are about the size and shape of a credit card. There are three types of PC cards for adding additional memory (RAM and hard disks) or connecting peripherals (such as network devices) to your computer. Popular with laptop computer users.

PCU Printer Control Unit, an IBM term.

PDL A Page Description Language is a clever short-cut for transmitting bitmapped images from a PC application to a printer. They save processing time, by sending only "instructions" to a printer, rather than the entire bitmapped image. They also allow the printer to print any font, any size. PDL is the generic term. Hewlett-Packard has been the major proponent of PDLs. They include one called PCL — Printer Command Language — with all their printers. Also "Print Description Language," which is what Xerox calls it.

Peek and Poke Viewing and altering a byte of memory by referencing a specific memory address. Peek displays the contents; poke changes it.

Moore's Imaging Dictionary

PEL Short for "picture element." What IBM literature insists on calling a pixel, which is also an abbreviation for "picture element." The smallest graphic unit that can be displayed on a screen.

Pen plotter See plotter.

Pen Windows A Microsoft operating system for notebook-size computers that use a stylus instead of a keyboard as the input device.

Pen-based computing Entering data into a computer with an electronic "stylus" or pen, and a pad that accepts the stylus' input.

Pending queue A workflow term. The electronic resting place for partially processed items awaiting additional information.

Pentium In the fall of 1992, Intel adopted the name Pentium for its 80586 chip, its successor to the 80486. It introduced the Pentium formally in early April, 1993. The chip is capable of 112 million instructions per second and is 80% faster than the fastest 80486. It contains more than three million transistors and is said to be a superscalar chip, which means it can execute two instructions at a time.

Perfect The kind of binding with a squareback, as opposed to stapled. Uses trimmed edges held together by glue and wrapped with a separate cover.

Perfecting Printing both sides of a sheet in the same pass.

Perforator An instrument for the manual preparation of a perforated tape, on which telegraph signals are represented by holes punched in accordance with a predetermined code.

Peripheral Any hardware device attached to a computer, such as a hard drive, printer, scanner or CD-ROM player.

Persistence A way to overcome flicker in a CRT that has a slow "refreshing" rate. The phosphors remain glowing, or "persist," after they've been energized.

Personal computer PC. A computer for personal single-user use, as opposed to other types of computers — mainframes and minis — typically shared by many users.

Personal groups In Windows NT, icons that you have put into groups which are stored with your logon information. Each time you log on, your personal groups appear.

Personal video system A PC-based video system for Windows introduced by AT&T in conjunction with its NCR subsidiary in the spring of 1993.

Petabyte A petabyte is equal to 10 to the 15th. A petabyte is equal to 1,000 terabytes.

PGH Presentation Graphics Facility.

Phase A of a fax call Phase A is the first part of a fax machine's call process. It is the call establishment. It occurs when transmitting and receiving units connect over the phone line, recognizing one another as fax machines. This is the start of the handshaking procedure. See PHASE B.

Phase B of a fax call Phase B is the second part of a fax machine's call process. It is the premessage procedure, where the answering machine identifies itself, describing its capabilities in a burst of digital information packed in frames conforming to the HDLC standard. See PHASE C.

Phase C of a fax call Phase C is the third part of a fax machine's call process. It is the fax transmission portion of the operation. This step consists of two parts C1 and C2 which take place simultaneously. Phase C1 deals with synchronization, line monitoring and problem detection. Phase C2 includes data transmission. See PHASE D.

Phase D of a fax call Phase D is the fourth part of a fax machine's call process. This phase begins once a page has been transmitted. Both the sender and receiver revert to using HDLC packets as during Phase B. If the sender has further pages to transmit, it sends an MPS and Phase C recommences for the following page. See PHASE E.

Phase E of a fax call Phase E is the fifth part of a fax machine's call process. This phase is the call release portion. The side that transmitted last sends a DCN frame and hangs up without awaiting a response.

Phase change ink jet A form of ink-jet printing in which solid inks are liquefied and jetted onto the page, where the ink immediately resolidifies.

Phase change ink-jet can print on a wide range of paper types because the solid ink does not wick into paper to degrade the image.

Phase change recording A rewritable optical recording technique. A laser causes the medium to crystallize in a controlled way, reflecting light either into or away from the reading laser.

Phase change is a faster rewritable optical technology than magneto-optical because it directly overwrites old data in one pass. M-O requires an erase pass before there can be a new write. Phase-change discs are also less expensive than M-O.

Phonenet Farallon's twist on Apple's LocalTalk local area network. PhoneNet uses standard one pair UTP (unshielded twisted pair) wiring for networking. PhoneNet is compatible with LocalTalk.

Phoneme A voice recognition term. The minimal significant structural unit in the sound system of any language that can be used to distinguish one word from another. For example, the "p" of pit and the "b" of bit are considered two separate phonemes, while the "p" or spin is not. It's part of the "sp" phoneme. These minimal sound units comprise words.

Phonetic alphabet The authorized words used to identify letters in a message transmitted by radio or telephone: Alpha, Bravo, Charlie, Delta, Echo, Foxtrot, Golf, Hotel, India, Juliet, Kilo, Lima, Mike, November, Oscar, Papa, Quebec, Romeo, Sierra, Tango, Uniform, Victor, Whiskey, X-ray, Yankee, Zulu.

Phong A technique used in 3D illustration rendering that determines the brightest spot and the darkest spot in the image, and spreads out the remaining values. Faster than ray tracing, but doesn't create as accurate a scene.

Phosphor Substance which glows when struck by electrons. The back of a cathode ray tube face is coated with phosphor.

Photo CD A format for storing and retrieving color images on CD, developed by Kodak and Philips. Can also be used for data.

Photoconductor 1. Any transducer that produces a current which varies in accordance with the incident light energy. A fiber optic communications term.

2. Material, available in many forms (sheets, belts and drums), which changes in electrical conductivity when acted upon by light. Electrophotography (a form of facsimile machine printing) relies on

the action of light to selectively change the potential of a charged photoconductive surface, creating areas receptive to an oppositely charged toner, thus making the latent charged-image visible.

Photochromic Compounds that become dark when exposed to light, and can be made clear again by removing the light, or exposing them to light of another wavelength. Proposed as erasable optical storage media.

Photocomposition The manipulation and transfer of graphic images and text, using photographic means, to a light-sensitive paper or film.

Photodetector In a lightwave system, a device which turns pulses of light into bursts of electricity.

Photoelectric effect The emission of electrons by a material when it is exposed to light. Albert Einstein received a Nobel Prize for explaining this phenomenon. Amazingly, he never received one for his brilliant theories of relativity.

Photon The fundamental unit of light and other forms of electromagnetic energy. Photons are to optical fibers what electrons are to copper wires. Like electrons they have a wave motion.

Photooptic memory An optical storage technique that uses a laser to record data on photosensitive film.

Photosensor The light-sensitive reading device used in optical scanners.

Phototypesetter Device that uses photographic techniques to reproduce machine-readable text on light-sensitive paper and film.

Photovoltaic Using light to produce electricity. Shine light on a device, typically a "cell." If the device produces electricity, that's called the photovoltaic effect. It's not very efficient at present. Less than 10% of the light energy emerges as electricity. But it's getting better.

Pi characters Special non-text characters, such as mathematical symbols. They are so rarely used they are not included among the special characters provided in a normal font.

Pica Unit of measurement used in typography and graphic design. Approximately 1/6 inch. Currently, in most desktop publishing systems, a pica is defined as exactly 1/6 inch. To be precise, a pica equals 12 points.

PICK A computer operating system written by VMark Computer Inc. Pick is very neat, but it never really caught on.

Picker In jukeboxes, the "hand" on the robotic "arm" that grasps and moves the disc to or from the storage slot, disc drive or mailslot. More boringly called the "carriage." Some jukes are "dual-picker."

Pico Prefix meaning one-trillionth.

Picosecond One-millionth of a millionth of a second. Nano is a billionth of a second, (or, for those in England, Australia or New Zealand, one thousand millionth of a second).

PICT A Macintosh bitmapped (raster) image file format.

PICT2 An expanded version of PICT.

Picture element Pixel, or pel.

PIECE Productivity, Information, Education, Creativity, Entertainment. Microsoft's trick for remembering the big five multimedia computing applications.

Piesio The Greek prefix meaning near.

PIF Progran Information File. A file that provides information about how Windows NT should run a non-Windows NT application. PIFs contain such items as the name of the file, a start-up directory and multitasking options for applications running in 386 enhanced mode.

PIM number Personal Identification Number. A group of characters entered as a secret code to gain access to a computer system.

Pinch roller The rubber-like wheel that presses mag tape against the drive's moving capstan to move it over the read/write heads at the proper speed.

Pincushioning When a video screen is distorted — with the top, bottom and sides pushing in — the screen is said to be suffering pincushion distortion.

Pinouts The purposes of each of the wires (pins) in a multiline connector.

Pit Broadly used now for all the data-carrying marks in optical discs. Originally meant the rimless troughs in the photo-resist layer of optical disc masters. Pits and flats (the unmarked areas) represent data.

Pitch 1. The number of characters per inch measured horizontally. Fixed spacing printers have the same pitch for every letter, regardless of the letters' widths. Proportional spacing has varying pitch, depending on the letter.

2. The distance between grooves (measured center to center) on an optical disc.

Pixel A sort-of acronym for Picture Element. Also called a Pel. When an image is defined by many tiny dots, those dots are pixels. On the printed page, each pixel is one dot. On color monitors, though, a pixel can be made up of several dots, with the color of the pixel depending on which dots are illuminated, and how brightly.

Pixel whackers Graphic artists.

Planar board IBM's new name for a motherboard in their System/2 Personal Computers. A motherboard is the main board in a PC on which the main CPU, the main memory, the clock and sundry other things like serial and parallel ports are mounted. Other boards, i.e. graphics boards, are plugged into the motherboard. Thus the expression "motherboard." No one knows why IBM dropped the word. Maybe it was too risque? Maybe they included more on their motherboards in the System/2 series that they would no longer function as motherboards? Maybe a feminist group of mothers objected?

Planetary camera A microform camera system in which the document is held still on a copyboard while film is being exposed. Once the document is recorded, it is replaced with the next document to be filmed, and the film in the camera is advanced. More accurate, but slower, than a rotary camera.

Plasma display A display screen that works by energizing a gas sandwiched between two panels. The panels are divided up into an array of dots; to illuminate a certain area all the dots in its location in the x-axis (horizontal) and y-axis (vertically) are charged with electricity. This makes the gas in those areas glow. They are great displays, but quite expensive.

Plat When a CAD/CAM plotter prints something, the drawing is called a plat.

Platemaking The act of exposing a printing plate through a negative or a flat of negatives.

Platen The cylinder in impact printers and typewriters around which the paper goes and which the printing mechanism strikes to produce an impression.

Platform A loosely applied word for a software operating system and/or open hardware, which an outsider could write software for.

Platform independence A term from IBM and Metaphor Computer Systems. The idea, they say, is to produce a layer of software that would rest atop any operating system on any piece of hardware. The applications developers would write their software just once, rather than start from scratch each time they wanted get their software working on a different computer. If the whole idea sounds rather daunting, you're right.

Platter Another word for optical disc. Also the circular recording parts of a hard disk.

Playback A multimedia term. Playback is the process of viewing multimedia materials created by an author. Playback can include a range of activities, from viewing a single video clip to participating in a series of interactive multimedia training modules. Some playback applications (for example many training and presentation applications) are sold separately from their authoring applications. However, many developers are selling authoring and playback capabilities in a single product.

Playback head The part which converts the magnetic information on the tape or disk into an electrical signal. Moving the magnetic fields on the medium (tape or disk) past the playback head generates a tiny voltage, which is picked up in a conductor (a coil) in the playback head and sent onto the electronic equipment where it is amplified or transmitted.

Plot To use vector graphics; that is, to draw images with many straight lines, rather than dots.

Plotter A printer that printed vector graphics, i.e., images created by a series of many straight lines. Uses pens on robotic arms, sometimes with different colored inks.

Plotter font A font created in vector graphics, i.e. a series of end points connected by lines. Plotter fonts can be scaled to any size and are most often printed on plotters. Some dot-matrix printers also support plotter fonts.

Plug 'n play The fantasy that a piece of equipment can be installed and work instantly, with no trouble. AKA a lie.

PLV Production Level Video. DVI Technology's highest quality motion video compression algorithm. It's about 120-1 compression. Compression is done "off-line". i.e. non-real time, and playback (decompression) is real time. Off-line compression will always produce a better image quality than real time since more time and processing power is used per frame.

PMF Print Management Facility. How IBM controls page printers from mainframes.

PMMA Polymethylmethacrylate, better known as Plexiglas or Lucite. Many video discs are made from it. Tends to absorb moisture, but otherwise perfect for the job.

PMS Pantone Matching System. A de facto standard color language. A means of describing colors by assigning them numbers.

POGO Post Office Goes Obsolete. When MCI Mail was being planned, its code name was POGO. Cute.

Point Unit of measurement in typography, approximately 1/72 inch. Currently, in most desktop publishing systems, a point is defined as exactly 1/72 inch. See pica.

Point size Same as type size or font size.

Pointer A logical link from a database record to another record, file or image stored elsewhere.

Polygon One of the basic shape-making tools in a drawing program, along with circle, rectangle, arc, line. A polygon is many sided, its shape comprised of line segments between points. It shape can usually be altered by "dragging" any of the convergence points.

POND Print ON Demand.

Port The channel in a computer used for input and output to a peripheral device, such as a printer, monitor or modem. The most common ports are serial (COM) and parallel (LPT). Serial ports are used for devices that accept information one bit at a time. Parallel ports are used for devices that accept information eight bits at a time and are generally faster than serial ports.

Portable 1. A computer that you can carry and runs off its own power source. AKA laptop, notebook, palmtop. 2. Software that can be moved to and work on a different operating system or platform.

Portrait Page or monitor orientation in which the page height exceeds the page width. Contrast with landscape. Some monitors can change orientation.

Portrait mode In facsimile, scanning lines across the shorter dimension of a rectangular original. CCITT Group 1, 2 and 3 facsimile machines use portrait mode.

Positional index A full-text index that can identify both the presence of a word in a document, AND its exact location in the document.

Positive Photographic print that accurately represents the tone values of the original. Contrast with negative.

Posterization Video effect which reduces the range of color variation in a video image so that it looks flat or two-dimensional, like a paint-by-number picture.

PostScript A software published by Adobe Systems that translates graphics created in a computer to language a (PostScript-compatible) printer can understand. It's called a page description language. PostScript-compatible printers have interpreters in them that create the proper dot patterns to recreate the screen image — text and graphics — to a page of paper based on a PostScript file. A PostScript file is text, like any program, and can be edited at the text level. But you have to know what you're doing. Our art director, Saul Roldan, is presently teaching himself to edit PostScript. He has my best wishes.

The big advantage of PostScript is that it is device independent. Thus if you create an image (text and/or photo and/or drawing), you can print it to a relatively cheap, low-quality printer like a laser printer or a prepress-quality imagesetter, like a Linotronics.

Powerbook Apple's name for a line of laptop Macintosh computers it introduced in late 1991.

Moore's Imaging Dictionary

ppm Pages per minute. A measurement of the throughput speed of a scanner — how many letter-size pages the scanner can scan in one minute. Beware: ppm can be misleading. It may or may not include the time required to remove one document and place the next on the scanner. It may or may not include software processing time.

Pre-emptive, real-time support When Microsoft announced its At Work operating system, it said it had a number of key features, one of which was "pre-emptive, real-time support." Here's Microsoft's definition: Communication devices such as fax machines and phones are distinct from personal computers in that they have critical real-time needs. Consequently, the software in these devices must attend to communication hardware such as modems very frequently, so that pieces of the communication are not lost. To support this need, the operating system was designed to be able to put other processes "on hold" temporarily in order to service the communication hardware before continuing other functions.

Preferences Settings in an application that you adjust to your personal taste. For instance, desktop publishing packages can compute measurements in either picas or inches — your preference. Usually, preferences can be set, so you don't have to reset them each time you open the application.

Preferred term When several words mean the same thing, the most commonly used one is the preferred term. Useful in text-retrieval products that can locate documents based on synonyms of the word used in the search query.

Pre-fetching. See Pre-loading.

Pre-loading Or pre-fetching. Documents are manually (by the administrator) retrieved from a jukebox or optical drive in anticipation of some special need. Pre-loading takes advantage of the regular cycles of business, such as payroll processing on the 10th and 20th, personnel reviews the first week of each quarter, year-end processing of accounting reports. The imaging system administrator can take advantage of regular cycles to improve the system's response to users.

Premastering The phase of CD-ROM production in which machine-readable data is converted to CD format. At this stage, error correction blocks (of 288 bytes each) are added to every user block (2,048 bytes).

Moore's Imaging Dictionary

Prepress The preparation work required to turn "camera-ready" artwork into the printing plates needed for mass production, i.e., making negatives, "stripping" or placing the negatives in place, and etching the plates.

Pressure-sensitive pad A digitizing tablet that can measure the amount of pressure placed on the stylus, as well as its location. Used in paint packages to great effect.

Prewrite defect A defect in the groove of a pre-grooved optical medium. Measure in ratio to the available total recording time on the disc.

Primary partition A portion of a physical disk that can be marked for use by an operating system. In Windows NT there can be up to four primary partitions (or up to three, if there is an extended partition) per physical disk. A primary partition cannot be subpartitioned.

Primitives Standard geometric 3-D objects that a program can generate for the user. Examples are spheres, cubes and cones. Some programs only allow the user scale these objects and not to edit them.

Print engine Today's raster image printers (laser printers and imagesetters) have two basic functions: to translate either PDL data or bitmap data into an arrangement of dots that the printer is able to produce (called raster image processing). This is done by the RIP engine. The second function is to control the mechanism that actually lasers or inkjets the dots onto the paper. This processing is done by the print engine.

Print screen The command that instructs a computer to send the image presently on the screen to a printer.

Print server A shared computer in a network dedicated to queuing and sending printer output from all the other networked computers to a printer. The server will sort jobs by priority (the boss's letters get out first, form letters out last) and will queue users' requests, while they get on with their work. A print server allows users on a LAN to share an expensive printer.

Print spooler A separate "cache" of memory dedicated to accepting and sending print data to a printer, allowing the user to run another program without interference. There are two primary advantages to print spooling: 1. You can use the spooler to save your and your computer's time. Dump the report to a print spooler at thousands of bits per second. Get on with something else on the computer. 2. You can use a print spooler to schedule several users' printing requests. This is particularly good in multi-user environments — for example, where the printer is a laser printer (and therefore expensive) and is attached to a LAN (Local Area Network).

Printer There are many devices that translate machine language of one kind or another into physical hard copy. The evolution has been roughly: impact printers begat dot matrix printers begat thermal printers begat laser printers begat thermal wax printers begat dye sublimation printers.

Printer driver A program that controls how your computer and printer interact. For example, a printer driver file supplies Windows with information such as the printing interface, description of fonts and features of the installed printer.

Printer font A font stored in your printer's memory, or soft fonts that are sent to your printer before a document is printed.

Priority routing/queuing In a worklow, items can be tagged for priority processing. Some systems offer automatic, event-driven priority alternation (for example; after 2 PM, all orders in the queue of over $10,000 become top priority so they can be processed before the end of the day).

Privileges The access rights to a directory, file or program on a LAN or over a remote link. Typically read, write, delete, create and execute.

Pro750 One of Intel's Digital Video Interface (DVI) product families, introduced in 1989, consisting of Application Development Platform, boards and software.

Process color printing The recreation of color by combining two or more of the subtractive colors — cyan, magenta and yellow, plus black.

Process separation Creation of the negatives necessary to for process color printing. Publishing software automatically creates separate files which represent the correct tone values.

Processor The intelligent central element of a computer or other information-handling system.

Program file A file that starts an application or program. A program file has an .EXE, .PIF, .COM, or .BAT filename extension. AKA executable.

Programmer's keys The plastic keys that come with some Macintoshes that let you reboot without turning the power off. Macs do not have any other physical re-start buttons. You can restart through software.

Progressive proofs Also called progs. Color prints of each of the process colors negatives, layered in register to show what the final printed piece will look like.

P-ROM Partial ROM. An optical disc medium made by Verbatim that has magneto-optical and write-once sectors on the same 3 1/2" disc.

Proportional font A font in which different characters have varying widths. All magazines and newspapers are printed in proportional characters, which make reading easier. By contrast, in a monospaced font, such as one on an old typewriter, all characters have the same widths.

Proportional spacing Printers with proportional spacing allow the horizontal space, called the "pitch," to vary depending on the letters' width. Opposite of monospacing.

Prosody The intonations of normal speech. In text-to-speech conversion, prosody refers to how natural it sounds — the ups and downs of the sentence.

Protocol Any defined hardware or software methodology that, if adhered to, allows two devices to interoperate. In data communications, it refers to the format and timing between two devices. The protocols for data communications cover such things as framing, error handling, transparency and line control. There are three basic types of protocol: character-oriented, byte-oriented and bit-oriented.

Protocols break a file into equal parts called blocks or packets. These packets are sent and the receiving computer checks the arriving packet and sends an acknowledgement (ACK) back to the sending computer. Because modems use phone lines to transfer data, noise or interference on the line will often mess up the block. The purpose of a protocol is to set up a mathematical way of measuring if the block came through accurately. And if it didn't, to ask the distant end to re-transmit the block until it gets it right.

Protocol conversion A data communications procedure which permits computers operating with different protocols to communicate with each other. It requires a device, a protocol converter, to translate a binary data stream from one format to another, according to a fixed algorithm.

Protocol stack A group of drivers that work together to span the layers in the network protocol hierarchy.

Proximity search A feature of full-text searching, in which every occurrence of a word within a a certain distance of another word is found. I.e., finding every time the word "budget" is mentioned within 20 words of the word "Congress."

PS/2 IBM's current family of microcomputers which sport one major difference from its predecessors, namely a 32-bit Micro Channel bus. This "bus" moves information from the printed circuit cards and to other printed circuit cards, which may contain their own individual microprocessors (computers on a chip) and which may communicate with the outside world through their own communications ports.

PTR Abbreviation of printer.

PU Physical unit. A Microsoft definition: A network-addressable unit that provides the services needed to use and manage a particular device, such as a communications link device. A PU is implemented with a combination of hardware, software, and microcode.

Puck An input device used to create graphics, like a mouse but equipped with a number of special-function pushbuttons.

Moore's Imaging Dictionary

Q The 17th letter of the alphabet. In ASCII, uppercase "Q" is represented as hexadecimal 51; a lowercase "q" is hexadecimal 71. In EBCDIC, an uppercase "Q" is hexadecimal D8; a lowercase "q" is hexadecimal 98.

Q factor In JPEG compression, the amount of data that is "lost" is adjustable. When JPEG removes data from an image, it does not do so arbitrarily. The criteria are set by the person implementing it. This is called setting the "Q factor." Higher quality equals lower compression, and vice versa.

In general, vendors implement JPEG to first remove the data least likely to be seen by the human eye. Still, implementations are slightly different. As a result, the same Q setting for different vendors' JPEG produces different results. In the future, it is conceivable that vendors will tailor their implementation to handle certain images with greater compression and less loss.

Setting the Q factor is a matter of experimentation and differs for every image. There are no rules, at least not yet. No one has enough experience.

QBE Query By Example. A database front-end that requests the user to supply an example of the type of data to be retrieved.

QBF 1. The test message "The quick brown fox jumps over the lazy dog." It contains every letter of the alphabet. 2. Query By Form.

Quad In typesetting, aligning text in the center, to the left or to the right by adding spaces. It's really another word for alignment — quad left means flush left; quad right means flush right; quad center means centered.

QuarkXpress A high-end page layout program for Macintosh computers; renowned for its powerful typography tools. Used to create this glossary, and the magazines which Telecom Library also publishes (call 1-800-LIBRARY for more information).

Quark Publishing System. A just-released (summer 93) workflow extension to QuarkXpress that manages copy flow, editing, approval, etc. clear through the production process.

Quality Not as subjective as you think. Usually determined by technical specs. For example, the quality of an output device is measured in dots per inch and pages per minute. The quality of a monitor is measured by maximum resolution, scan frequency and dot pitch.

Quantizing The second stage of pulse code modulation (PCM). When the waveform samples are measured to find their precise amplitude, they are quantized. These measurements are then converted to binary code, or "digitized." Also called sampling.

Quantizing Noise The inability of an analog signal to be exactly replicated in digital form.

Query In a document management system, how you ask to retrieve a document. You usually query by keyword or by text appearing in the document's associated database.

Query by example The ability to use an existing set of text as the basis for a text search.

Query by form A database and text-retrieval front end, where the user is prompted to fill in blanks in order to build a query.

Query language A predefined set of rules and syntax that you must use to properly retrieve information from a database. Structured Query Language (SQL) is an attempt to define a standard query language. It's catching on.

Queue A stream of tasks waiting in line to be executed. The tasks — retrieval requests and write attempts to a jukebox, say — are usually logically managed for optimum performance. See queue management.

Queue management In a network, tasks like retrieval and writes to a jukebox come randomly from all the users. These tasks vary in urgency — retrievals are higher priority than writes, for example. Queue management sorts out requests from the network by priority.

Queue management also enhances the performance of a jukebox, by intelligently re-ordering requests. For example, if there are three requests for images on platter 1 and two from platter 2 and the another from platter 1, queue management means the requests from platter 1 will get handled together, then go to platter two. Sometimes it's called "elevator sorting" — responding to requests in logical order, not in the order in which they were made.

QuickTime Apple's standard file format for computer animations. Includes editing tools and programming that supports software efforts in playback.

Quiscient A fancy word for quiet. No noise. No activity. Doesn't exist in New York City, where this dictionary was written.

Quiet zone The area on the edge of the page where there is no printing, i.e. the margin.

QWERTY keyboard Common keyboard arrangement of the characters, digits and punctuation marks, named after the sequence of characters near its upper left corner. Computer keyboards add special keys, such as function keys, control keys, delete, insert, etc., which didn't exist when the QWERTY keyboard was designed. The layouts of computer keyboards therefore vary enormously. There is no "standard" keyboard. Contrast with Dvorak keyboard.

R The 18th letter of the alphabet. In ASCII, uppercase "R" is represented as hexadecimal 52; a lowercase "r" is hexadecimal 72. In EBCDIC, an uppercase "R" is hexadecimal D9; a lowercase "r" is hexadecimal 99.

Radial acceleration The rate at which a track on an optical disc accelerates toward and away from the center, because it is not perfectly aligned or perfectly round.

Radio frequency That group of electromagnetic energy whose wavelengths are between the audio and the light range. Electromagnetic waves transmitted usually are between 500 KHz and 300 GHz. Also known as RF.

Ragged Typesetting so that one side of the column is not aligned.

RAID Redundant Array of Inexpensive Disks. A storage device that uses several hard disks working in tandem to increase bandwidth output and/or to provide fault tolerance and backup. Data segments are either distributed across all the disks, or duplicated on each disk, depending on which "level," or implementation of RAID, you use. These "levels" are not set in concrete and vendors offering compatibility with a certain level aren't. Levels 1 and 5 are the most popular.

RAID Levels Level 1 mirrors data — all data is written on two or more separate disk drives for redundant back up.

Level 2 uses one or more disks to encode an error checking code for all the other disks. You get more storage space than Level 1, and the error correction provides reliability.

Level 3 interleaves data across all the disks, and uses one disk for parity checking (error correction on the fly). Provides high transfer rates.

Level 4 is similar to Level 3, but interleaves data only in "chunks," or sectors, 512 bytes long. Also has a dedicated error-correction disk.

Level 5 has no dedicated error-correcting disk. It does parity checking on the data on all the disks. Highest storage level of them all.

RAM Random Access Memory. The primary memory in a computer used by applications to perform tasks while the computer is on. Memory that can be overwritten with new information. The "random access" part of its name comes from the fact that all information in RAM can be located — no matter where it is — in an equal amount of time. This means that access to and from RAM memory is extraordinarily fast. By contrast, other storage media use "sequential access memory," whose access time varies according to the location of the data on the medium. When you turn the computer off, all information in RAM is lost.

RAMDAC Random Access Memory Digital-to-Analog Converter. The chip on a VGA board that translates the digital representation of a pixel into the analog information needed for display on the monitor.

RAM drive A portion of memory that is used as if it were a hard disk drive. RAM drives are much faster than hard disks because your computer can read information faster from memory than from a hard disk. However, information on a RAM drive is lost when you turn off or restart your computer. Also known as virtual drive or RAM disk.

Ramp In graphics, the consistency with which a gradated color flows from one to the other. Good programs allow you to adjust the ramp, and save it as a template for future use. Example: The color changes from red to blue. With a standard ramp, the left would be red, the right blue and the dead center would be purple. You should be able to adjust the ramp, so the change to blue happens off center, toward the right.

Random storage Direct access storage devices, such as hard drives and floppies, write data anywhere on the disk that is free and quick to get to. They even break up contiguous files into scattered chunks of data all over the disk. The whereabouts of all these chucks is registered on a file allocation table (FAT). Random storage is much faster to retrieve from than linear storage, such as that used in magnetic tape.

RAS Rodent Abatement System. In fall of 1992, NASA's brand new headquarters in Washington was plagued with rats. The agency, like all good agencies, made up an acronym for their countermeasure plan.

Raster Images that are stored and displayed in bitmap (pixels in horizontal and vertical rows) form. Contrast with vector.

Raster display The most common type of display terminal. Uses pixels in a column-and-row array to display test and images.

Raster font A set of characters for screen display or printing that is stored as bitmaps in specific character sizes. Compare with vector fonts.

Raster image processor RIP. The software and hardware that translates a page description language (PDL) like PostScript, or graphics output into the dots to be used by an output device's print engine.

Raster graphics Images defined as a set of pixels or dots in a column-and-row format.

Raw footage Video that you've shot (produced) but not yet edited (post-produced).

Ray tracing The technique for creating reflections and shadows in computer graphics for more realistic, 3-D effect. Takes the location of the "light source" and computes where the light rays would fall.

RC paper Resin-coated photographic paper; commonly used for imager output.

Read To retrieve information from a storage device, usually meaning a disk or disc. A read head locates the sector of data, senses the magnetic biases of the substrate (that's the reading part) and translates the pulses to the disk controller, and on to the processor.

Read-only file A file that you can read but cannot make changes to.

193

Read write cycle Time of reading and writing data onto a memory device.

Reconditioning Periodically fast-forwarding and rewinding a tape to retension it on the reel. Prolongs tape life and keeps data readable. Recommended.

Record (Noun, pronounced RE-cord) In a database, a record is a group of related data items treated as one unit of information — for example, policyholder's name, address, social security number, etc. Each item in the record is called a field. (Verb, pronounced re-CORD) To place information on a storage medium.

Record head The electromagnetic device which magnetizes the surface of a magnetic recording — tape, disk, etc. — in proportion to an electrical signal.

Record length The number of bytes in a record. See RECORD.

Recorder In video, the receiving tape or deck. The "to" or destination point in a video editing system. The machine you use to compile an edited video.

Recording zone The ring-shaped area of an optical disc on which data can be recorded.

Recto A right-hand page. Always odd numbered. Opposite of verso, or left-hand page.

Redline To append comments to a document and to make alterations with symbols — circle something, strike-through, underline. Many electronic document programs allow you append redline comments.

Reduce In Windows, to minimize a window to an icon at the bottom of the desktop, using the Minimize button or the Minimize command. A minimized application continues running, and you can select the icon to make it the active application.

Reduction rate The amount of compression and the resulting free disk space.

Reference track A special magnetic track placed on floptical diskettes used by the drive to calibrate the optical tracking system with respect to the magnetic recording tracks.

Reengineering What many big companies face when they think they're just "implementing imaging." The full benefit of imaging and workflow usually are gained only by redesigning basic business processes to meet new objectives — the main objective usually being to treat your customers better. There are three kinds of reengineering: 1. Life Cycle Reengineering. This is ongoing, incremental changes day-by-day. Tweaking what you got. 2. Crisis Reengineering. Changing to respond to pressure from upper management. Usually results in some process changes, but often no substantive change. 3. Goal-oriented re-engineering. Radical. Totally new, clearly defined objectives — very different from today's business objectives — are set. Requires commitment from all levels in your organization and a great deal of involvement from the "platform" vendor. That being, increasingly these days, the imaging/workflow vendor. The folks at Delphi Consulting Group are workflow experts: 617-247-1511.

The term was probably invented by Michael Hammer in the July-August, 1990, issue of Harvard Business Review. In that issue, he wrote "It is time to stop paving the cowpaths. Instead of embedding outdated processes in silicon and software, we should obliterate them and start over. We should 're-engineer' our businesses; use the power of modern information technology to radically redesign our business processes in order to achieve dramatic improvements in their performance."

Reflective art Artwork whose image does not appear by light shining through it or emitting from it, but is only visible because light is reflected from it. In other words, a picture versus a slide. (A slide is transparent art.)

Reflective read The type of optical medium where the laser to be read is reflected from the medium.

Refresh The phosphors at each pixel of a CRT which are stimulated by a charge from an electron gun glow only briefly. The must be renewed frequently in order for the image to appear stable. This renewal is called refreshing.

Refresh rate Measure of how often the image on a CRT is redrawn; expressed in hertz, or cycles. Above 70 Hz is considered flicker-free. The higher the better, though, because the mind is affected by flicker that the eye can't perceive.

Registration The precision with which CMYK primary colors are placed on the page relative to each other. Misregistration results in color inaccuracies, unclear edges and reduced image clarity.

Relational database The data adheres to a relational data model, so data can be stored in table-like structures called relations. The idea is that anyone using the right language can query — retrieve data from — the database regardless of the application used to construct it. SQL — Structured Query Language — is the de facto standard query language.

Relevance ranking Displaying retrieved documents in descending order of the closeness to which they match the query statement.

Removable media Diskettes or cartridges that can be removed from a computer drive. For example, a Bernoulli box uses removable cartridges.

Render In graphics, render is another word for "draw" or "illustrate." In 3D graphics, it means letting the software create the shadows and colors to build a 3D image.

Rendezvous In a parallel workflow, the point at which routes are brought together.

Resolution Indicates the number of dots that make up an image on a screen or printer. The larger the number of dots, and thus the higher the resolution, the finer and smoother images can appear when displayed at a given size. Low resolution causes jagged characters. The ideal resolution is a trade-off between quality, and the overhead in storage power and processing strength required to use it.

Restore Button In Windows, the buttons containing up and down arrows at the right of the title bar. The Restore button appears only after you have enlarged a window to its maximum size. Mouse users can click the Restore button to return the window to its previous size. Keyboard users can use the Restore command on the Control menu.

Resume Enables laptop computer to return to the place you left off working without rebooting. Do not rely on "resume." It seems to know exactly when to crap out.

Retired sectors Sectors on a disk or disc which contain errors. The defect management software marks them as "retired."

Retirement The term that describes the decision to throw away a recording medium (optical disc or mag tape) when it has too many defects.

Retrieval key A word, number or phrase associated with a document to aid in its retrieval from storage. Sometimes called descriptors. There are often many retrieval keys used together to fully locate a document; together they are called an index.

Retensioning Periodically fast-forwarding and rewinding a tape on the reel. Prolongs tape life and keeps data readable. Recommended.

Reverse video Display in which characters are highlighted by inverting the normal display mode; for example, if characters are normally bright against a dark background, a block of reverse video text would be displayed as dark letters in a bright rectangle.

Rewritable optical Optical media from which data can erased and new data added. Magneto-optical, phase change and dye polymer are the types of rewritable optical discs.

RGB Red, Green, Blue. The primary colors, called "additive" colors, used by color monitor displays and TVs. The combination and intensities of these three colors can represent the whole spectrum.

RIFF Resource Interchange File Format. Platform-independent multimedia spec from Microsoft. Allows audio, images, animation and other multimedia elements to be stored in a common format.

Right reading Normal, left-to-right image reproduction. Contrast with wrong reading.

RIP Raster Image Processor. The software and hardware that translates bitmap and PDL instructions from a graphics program to the print mechanism.

RISC Reduced Instruction Set Computing. A computer system with a special microprocessor that processes fewer instructions, and thereby is much faster. A RISC system depends on software to perform many of the functions that would normally be done by microprocessors. RISC workstations are used in calculation-intensive operations such as those performed by computer-aided design (CAD) and computer-aided manufacture (CAM) engineers.

River A typesetting error where spaces align on contiguous lines to form a distracting "river" of white space.

Moore's Imaging Dictionary

RLE Run Length Encoding. A compression algorithm used in Group III (fax-standard) compression. A scanner reads a horizontal line of pixels — the "scan line" — and counts the black pixels. This is the "run length" or "run count." The same is then done for white pixels. A scan line might have a run of 230 white pixels, followed by a run of 23 black, followed by 45 white, etc. The result is a set of run lengths for the line. That's RLE. The different run lengths are then given codes — shorter ones for the most frequently occurring run lengths, longer ones for the unique run lengths — to represent them. That's called "Huffman Coding." The codes are set in Group III compression; some compression schemes can assign new codes for each image.

ROFL Abbreviation for "Rolling On the Floor, Laughing;" commonly used on E-mail and BBSs (Bulletin Board Systems).

ROM Read Only Memory. Data stored in a medium that allows it to be accessed but not erased or altered. The contents of ROM stay after the computer is turned off.

ROM font A PC's type font. It consists of a set of 256 characters which cannot be edited — unless you are running in video mode, in which case you can design your own type font.

Roman Regular, non-italic non-bold type in any font.

Rosette The visible pattern of the four-color printing dots on paper or proofing material. Actually caused by a film creation system that is TOO accurate. Solved by fudging the screen angles or registration by a half-dot or so.

Rotary camera A microform camera system in which documents are recorded "on the fly;" i.e., the documents are fed and the film is advanced simultaneously by synchronized transport systems. Rotary cameras are faster than their planetary counterparts, but because of the vibrations from the film and document motion, their resolution is lower.

Rotational latency The time required for a disk or disc to rotate under the read/write head until the correct sector comes around.

Rough-cut A loosely compiled videotape of footage you want in your finished production but without the final editing or transitions.

RS-232-C One of the communmications ports that come with most

PCs. It's also called the serial port. Modems and some printers connect to PCs through the RS-232-C port.

RTV Real-Time Video. DVI software that implements quick-and-dirty, realtime video compression. Once called "edit-level video," it stores video as only 10 frames per second. Meant for use while developing DVI applications.

Run Length Encoding. A compression algorithm. See RLE.

Runout Deviation from the perfect motion. For instance, if a horizontal spinning disc is slightly warped, it wobbles up and down — that's axial runout. A disc with an off-center spindle will move back and forth — that's radial runout.

Moore's Imaging Dictionary

S The 19th letter of the alphabet. In ASCII, uppercase "S" is represented as hexadecimal 53; a lowercase "s" is hexadecimal 73. In EBCDIC, an uppercase "S" is hexadecimal E2; a lowercase "s" is hexadecimal A2.

SAA Systems Application Architecture. A set of specifications written by IBM describing how users should interface with applications and communications programs. The idea is to give all software "a common feel" so that training will be less burdensome.

Saddle stitch The type of binding that uses staples (actually short bits of continuous wire) to hold pages together. So called because the groups of pages, or signatures, straddle each other saddle-style.

Safe title area The central portion of a video frame where you can place text and graphics and be sure no part will extend beyond the visible portion of the screen when viewed on some TVs or monitors.

Sampling Converting continuous analog signals, like voice or video, into discrete values, e.g. digital signals. Also called quantizing.

Sampling rate The number of times per second that an analog signal is measured and converted to a binary number — the purpose being to convert the analog signal into its digital analog. The most common digital signal — PCM — samples voice 8,000 times a minute.

Sans serif "Without a serif." Type fonts which do not have the short cross strokes at the top and bottoms of the main strokes. Sans serif are generally considered better for display and signage; serif fonts are better for pagefuls of text.

SAR Storage And Retrieval device. A tape changer that holds up to 300 rols of microfilm. Robotics and software retrieves the cartridge, locates the image and scans it for presentation to the user.

SASI Shugart Associates System Interface. The first SCSI interface specification defined by Shugart, a disk drive manufacturer. Later it was modified and renamed as the Small Computer System Interface (SCSI). See SCSI.

Saturated recording An optical recording device which causes the medium to stop absorbing light when the mark is fully formed. Useful because of the variation in laser intensity from power fluctuations and shifts in focus.

Saturation Refers to the amount of a color. Fully saturated colors are vivid, while colors which lack saturation look washed out or faded. Over-saturated color on video looks garish, and bleeds into adjacent areas.

SAW Surface Acoustic Wave. The newest type of touchscreen. It senses molecular waves traveling over a clear glass overlay.

Scale To Gray Improving document images by adding gray pixels to fill in jagged edges of characters which have been magnified to a larger size.

Scaling Process of uniformly changing the size of characters or graphics.

Scalable The ability of an application to grow easily and directly from a single-user version to a networked, multi-user version. Another usage: Microsoft refers to Windows NT as "scalable," because it runs on everything from Intel to RISC processors and single to multiprocessor systems.

Scalable typeface A font that can be enlarged or reduced to any size.

Scaled point size A font's point size that approximates a specified point size for use on the screen. For example, text that prints at 10 point on the printer may be represented by a slightly larger font on the screen to make up for the screen's lower resolution.

Moore's Imaging Dictionary

Scan To convert human-readable images into bit-mapped digital machine-readable code.

Scan head The part of the mechanism of a scanner that optically senses the text or graphic as it moves across a page.

Scan line The pixels that result from one sweep of a linear-type scanner, such as that in a fax machine.

Scan rate Number, measured in times per second, a scanner samples an image.

Scanner A device that optically senses a human-readable image, and contains software to convert the image to machine-readable code.

SCOOP Self COupled Optical Pickup. An optical drive design that combines the functions of the laser reading device with the photodetector used to accept tracking and focus error signals.

Screen Series of dots (may also be series of lines or other pattern) used to convert continuous tone artwork into a halftone. Resolution is measured in lines (of dots) per inch, or lpi.

Screen angles Angles at which halftone screens are placed to avoid moire patterns or to make up for poor reproduction. Common angles in four color process are: black = 45 degrees; magenta = 75 degrees; yellow = 90 degrees; cyan = 105 degrees.

Screen capture To transfer what presently appears on a display screen to a computer file. See print screen.

Screen dump Same as print screen. In a DOS machine, the command is shift-PrtSc.

Screen font A display-only type font that very nearly matches its companion printer font. Required for true WYSIWYG.

Screen saver A moving picture or pattern that appears on your screen when you have not moved the mouse or pressed a key for a specified period of time. Screen savers prevent screen damage — called phosphor burn-in — caused when the same areas of light and dark are displayed for long periods of time. My two favorites: Flying Toasters and the Lawnmower Man, "After Dark" screen savers by Berkeley Systems, 510-540-5535. What's your favorite? My phone is 212-691-8215; my fax is 212-691-1191.

Screening Converting a continuous tone photograph of artwork

Moore's Imaging Dictionary

into a halftone.

Scroll To move up or down through text or graphics in order to see parts of the file or image that cannot fit on the screen. Moving left or right is actually called panning.

Scroll arrow An arrow on either end of a scroll bar that you use to scroll through the contents of the window or list box.

Scroll Bar A bar that appears at the bottom and/or right edge of a window whose contents are not entirely visible. Each scroll bar contains a scroll box and two scroll arrows.

SCSI Small Computer System Interface. Pronounced "scuzzy." An industry standard (of sorts) for connecting peripheral devices and their controllers to a microprocessor. SCSI defines both hardware and software standards for communication between a host computer and a peripheral. Computers and peripheral devices designed to meet SCSI specifications should work together.

A single SCSI adapter card plugged into an internal IBM PS/2 micro channel PC slot can control as many as seven different hard disks, optical disks, tape drives and scanners — without siphoning power away from the computer's main processor.

SCSI was formerly known as SASI, for Shugart Associates Systems Interface, who invented it. The name was changed when it became a standard.

SCSI-2 A 16-bit implementation of the 8-bit SCSI bus. Using a superset of the SCSI commands, the SCSI-2 maintains downward compatibility with other standard SCSI devices while improving upon reliability and data throughput.

Scuzzy See SCSI.

SDLC Synchronous Data Link Control. A bit-oriented synchronous communications protocol developed by IBM where the message may contain any collection or sequence of bits without being mistaken for a control character.

SECAM Sequential Couleur Avec Memoire (sequential color with memory). A color television system with 625 lines per frame and 50 fields per second developed by France and the USSR. Color difference information is transmitted sequentially on alternate lines as an FM signal.

Secondary storage Another name for off-line. Refers to tape backups and transaction log recorders. Also used to create distribution copies.

Sector The smallest addressable unit of an optical disc's track. Contains 512 bytes.

Seek error The inability of an optical drive to find the user's request because of disc flaw or vibration or the drive just doesn't work right.

Seek time The time it takes to find some information on a disk (hard or soft) in a computer. Average seek time is a critical measure of the speed of a computer disk drive.

Select To mark an item so that an action can be carried out on that item. You usually select an item by clicking it with a mouse or pressing a key. After selecting an item, you choose the action that you want to affect the item.

Self indulgence The term for this author's habit of inserting irrelevant definitions in this dictionary for the sheer fun of it. Like this one.

Sensitive layer The layer in an optical medium where the data is recorded; it may be composed of more than one layer or materials. It is sandwiched by protective and supporting layers.

Sensitivity Measure of the light dose needed to mark an optical medium.

Sensor glove An interface device for experiencing virtual reality with the hand. Wired with sensors, it detects changes in finger, hand and arm movements and relays them to the computer, allowing users to manipulate and move things in a virtual environment.

Separations Color separations. The result of filtering a color imaging into is primary color elements. Used usually to refer to the film you get from a CMYK filtering process.

Sequencer A controller that uses MIDI interface commands to manipulate devices (usually musical instruments) in parallel or serial sequence.

Sequential access The technique of storing data in a contiguous line, one dit after another, as in a streaming tape device. Slower than random access, and less expensive.

Moore's Imaging Dictionary

Serial interface An interface between a computer and a serial device, such as a printer or modem, by which the computer sends single bits of information to the device, one after the other. Serial, asynchronous and RS232 interfaces are all the same type. Contrast with parallel.

Serial printer A printer using a serial interface, which you connect to a serial port.

Serial port A connection on a computer, labeled COM1, COM2, etc., where you plug in the cable for a serial device. Common serial devices are printers and modems. Windows supports COM1 through COM4.

Serif Type font which features tiny cross strokes at the tops and bottoms of the main strokes. Considered better for reading text. Sans serif (without serif) is better for display and signage.

Server A computer on a network which is dedicated to one task and usually shared by workstations (sometimes called clients). A database or directory server would be responsible for responding to a user's search request, returning the list of stored documents that meets with the parameters of the request. A printer server would take files from the clients and dole them out to an attached printer. An image server handles the storage and retrieval tasks for image access requests from the network. Etc.

Servo Short for servomechanism. Devices which constantly detect a variable, and adjust a mechanism to respond to changes. A servo might monitor optical signal strength bouncing back from a disc's surface, and adjust the position of the head to compensate.

Setup The black reference level of an input video signal. Adjusting the setup controls the level of black in your picture. For example, you could make black appear as dark gray instead of pure black.

SGML Standard Generalized Markup Language. An ISO standard text-description language (analogous to a page description language, such as PostScript) that translates text formatting features like bolding, columns, point sizes into a standard language for transfer between machines and applications.

Shadow mask The most common type of color picture tube in which the electron beam is directed through a perforated metal mask to the desired phosphor color element.

Shared screens A multimedia concept. Shared screen applications enable two or more workstations to display the same screen simultaneously. For example, two users sharing a screen can work on the same spreadsheet. Changes made by one user can be seen by the other as they are made. Shared screens can be implemented in two ways. One way enables people to view each other's screen while one person makes changes. The other way enables people to run the same application on both screens so that both users can make changes simultaneously.

Shared whiteboards A multimedia concept. Shared whiteboards enable you to "mark-up" a screen using a mouse or stylus input device and have the results show on other screens, often communicating over long distance telephone lines. The concept is similar to a traditional whiteboard mark-up process where everyone has a different color marking pen to circle, write or cross out items. The background board can be a window from the workstation such as a spreadsheet, image, or blank canvas, or it can be the entire workstation screen. The shared whiteboard can be used for either real-time or store-and-forward collaboration. In the store-and-forward scenario, the mark-ups can be implemented in a time-delayed fashion so everyone can follow the entire step-by-step process.

Shear A tool for distorting a selected area vertically or horizontally.

Sheet A piece of paper. Refers to both sides, although it can be any size or weight.

Sheetfed A scanner design in which the original document is transported past a stationary scanning head. Downside: subject to vibration and skewing. Upside: the fastest scanning technique, with the additional bonus of allowing duplex (both-side) scanning.

Sheetfeeder A device that holds many sheets of paper and delivers them individually into a scanner or printing device. Also called autofeeder or automatic document feeder — ADF.

Shilling fraction A way of typesetting fractions so that the numerator and denominator are aligned horizontally and both rest on the baseline.

Shortcut key A key or key combination you press to carry out a command or action. If a Windows command has a shortcut key, the key combination is listed to the right of the command name on the menu. For example, pressing the shortcut key ALT+F4 closes the active application.

Shrinkwrapped Software that you can buy from a store or catalog and install it yourself.

Signature A collection of pages printed together on one sheet, gang-style, to be folded and trimmed in such a way so that their order and orientation is correct. Also called a "form." Usually created in increments of eight.

Silver Common precious metal, the light-sensitive salts of which are used in photographic reproduction.

SIMM Single In-line Memory Module. Add-on memory chips that increase a computer's RAM. The most common SIMM is the 30-pin, 9-bit wide "1 by 9," which is the standard memory upgrade for PCs.

Simple device A device that adheres to the Media Control Interface (MCU) standard but does not use media files. An audio compact disc player is a simple device.

Site license You buy one copy of a software program and a license to reproduce it up to a certain number of times. Site licenses vary. Some require that a copy be bought for each potential user — the only purpose being to indicate the volume discount and keep tabs. Others allow for a copy to be placed on a network server but limit the number of users who can gain simultaneous access. This is called a concurrent site license. Many network administrators prefer this concurrent license, since it gives them greater control. For example, if the software is customized, it need be customized only once.

Skew 1. When a document is crooked when it's scanned. Skew causes OCR errors, and reduces legibility in document viewing. Many scanning softwares feature "deskewing" — the best ones use the text itself as a reference; others need a horizontal or vertical rule as a reference. 2. To slant a selected item in any direction; used in graphics and desktop publishing.

Slide Video editing transition effect where clip "A" moves off the screen as clip "B" moves onscreen to replace it. The video images actually move with respect to the screen (as opposed to a wipe). It's like using one slide to push another one out of your projector, except the images move during the transition.

207

Smart card A credit card-sized card which contains electronics, including a microprocessor, memory and a battery. These cards can be used to store data and are easily portable.

Small caps Typesetting all letters in capitals, but setting the normally lowercase letters in a smaller point size. Example: SMALL CAPS.

Smoke test If it smokes, it doesn't work!

SMPTE The Society of Motion Picture and Television Engineers developed a time code, based on a 24-hour clock, used to synchronize sound to moving pictures.

SNA Systems Network Architecture. IBM's very successful means of networking remotely located computers. The quickest definition is that it is tree-structured architecture, with a mainframe host computer acting as the network control center. Unlike the telephone network, which establish a physical path for each conversation, SNA establishes a logical path between network nodes, and it routes each message with addressing information contained in the protocol.

Soft boolean operators A fuzzy searching method in which the boolean operator AND is treated as OR. Resulting hits are ranked lower in relevance order than the hard use of AND.

Soft font A font, usually provided by a font vendor, that must be installed on your computer and sent to the printer before it can be printed. Also known as downloadable font.

Soft WORM An unappetizing name for an optical WORM technique that uses rewritable technology (such as M-O or phase change) but prevents you from altering data by placing logical write-protection (software code) on each disc.

Software The set of instructions that make computer hardware perform tasks. Programs, operating systems, device drivers and applications are all software.

Software-only video playback A multimedia term. Video software playback displays a stream of video without any specialized chips or boards. The playback is done through a software application. The video is usually compressed to minimize the storage space required.

Softwares I think it's OK to refer to software as plural when you mean "software programs." For example: "There are dozens of document management softwares under $200." Other people think it sounds stupid. Oh well.

Sound driver A software program that applications use to play sounds on your computer.

Soundex A play on "index." A search based on the phonetic sound of a query word, versus its spelling.

Source In video, the originating tape or deck. The "from" or origination point in a video editing system. Usually consists of rough-cut footage or computer-generated images.

Source document The document from which a linked object in OLE originates.

Spindle The center part of a disk (or disc) drive which maintains the axis of rotation and provides the force to rotate the disk (or disc).

Spin-up The time during which a drive accelerates its disk/disc up to operating speed.

Spin-down The time during which a drive decelerates to a stop.

Spline A mathematically defined curve that smoothly links a series of dots. See Bezier curve.

Split bar Divides a window into two parts. In File Manager, a directory window is divided by a split bar: The directory tree is on the left, and the contents of the current directory are on the right.

Splits In OCR, character bitmaps that are broken into pieces. Happens when the original document was degraded or light, or the scanner's threshold was set too high. Cause many OCR errors.

SPOOL Simultaneous Peripheral Operations On-Line. Another word for a queue; data is sent to a SPOOL to wait to be printed or stored.

Spooler The software that holds queued data waiting for the availability of a printer.

SQL Structured Query Language. A database query and programming sublanguage. Pronounced "sequel." It's an established set of statements used to add, delete or update information in a table, or request information from one or more tables in the form of a report.

Stacked fraction Typesetting a fraction so that the numerator and denominator are stacked vertically.

Staging In a document retrieval from an optical jukebox, the process where the image is fetched from the server by the software, and stored on the user's local PC until it is used. After it is used, it has to be manually deleted from the PC's disk (the "original" of the image is still in the jukebox), leaving magnetic disk space for other documents to be staged.

STAIRS Storage And Information Retrieval System) IBM's proprietary text document management system. Used on large mainframe computers, it allows users to search for documents based on key words or word combinations.

Standard Mode A windows operating mode that can be used with 80286, 80386 and 80486 computers. This mode provides access to extended memory and also enables you to switch between non-Windows applications, but it does not provide virtual memory or enable non-Windows applications to run in the background or in a window.

Static information In OCR, describes the general knowledge sources — dictionaries, syntax rules — that apply to all documents. See Dynamic information.

Static object An object that has been "pasted onto" a document, but not linked or embedded. Static objects cannot be changed from within the document. To change it, you must clear it from the destination document, change it with the application used to create it and paste it back into the destination document. To overcome this hassle is exactly why they invented OLE.

Status bar In Windows, a line of information usually located at the bottom of a window. In File Manager, the status bar shows the number of bytes available on the disk and the total disk capacity. Not all windows have a status bar.

STG Scale To Gray. Using gray pixels to fill in jagged edges of document images. STG improves readability. According to a study commissioned by Cornerstone and done by Dr. Jim Sheedy, the ability to read STG images was improved between 4% and 19%, depending on the resolution, and symptoms such as headaches, tired back, blurred vision were cut way down.

Moore's Imaging Dictionary

Stem The main vertical part of a character, such as the vertical stroke of a capital "T."

Sticky notes The ability of a document imaging system to allow annotations to scanned documents in text boxes. So-called because the text annotations are analogous to yellow Post-it notes.

Still-frame video A single frame of video which may be digitized and stored on your computer. Still-frame video clips don't have a fixed duration, so you can display the image for any length of time you want (see also freeze and grab).

Stitching Combining multiple strips of an image captured with a hand scanner into a single large image.

Stochastic screens Halftone screens, produced by imagesetter Raster Image Processors (RIPs), made up of randomly spaced equal-sized (small) dots instead of traditional grid-aligned dots. Supposed to reproduce better color from coarser line screens and create less moire.

Storage and retrieval system An imaging system implemented to allows users to search and gather large numbers of documents. See workflow system.

Storage media The physical device itself, onto which data is recorded. Mag tape, optical discs, floppy disks are all storage media.

Storyboard The video editing "worktable," where you put your video, audio and graphics clips into order, add video effects and transitions and do audio editing to create your finished video production.

Storyline Any of the parallel lines on a storyboard on which you arrange clips. Each media type (video, audio or graphics) gets its own storyline.

Streaming tape Storage medium; 1/4-inch magnetic tape used to store backup copies of files.

String A series of characters, usually the subject of a text search.

Strobe A video effect which gives the impression that an image is moving in fixed steps by displaying on every nth frame and repeating it n-1 times.

Stroke One line segment in a vector graphic image.

Stroke edge An OCR term. The line of discontinuity between a side of a stroke and the background, obtained by averaging, over the length of the stroke, the irregularities resulting from the printing and detecting processes.

Stroke speed In facsimile systems, the number of times per minute that a fixed line perpendicular to the direction of scanning is crossed in one direction by a scanning or recording spot. In most conventional mechanical systems, this is equivalent to drum speed. In systems in which the picture signal is used while scanning in both directions, the stroke speed is twice the above figure.

Stroke width In character recognition, the distance measured perpendicularly to the stroke centerline between the two stroke edges.

Style tags Prearranged codes placed in wordprocessed text which are interpreted by a desktop publishing program to adjust formatting, size, font, etc.

Stylus Pen-like instrument used with a graphics tablet as an input device.

Subdirectory A directory within another directory.

Submarining When you drag your cursor across a screen and the cursor disappears as you move it. Happens most on LCD screens because they change slowly — much slower than CRTs or VDTs (glass screens).

Subroutine A piece of a larger software program. Performs a specific task.

Subtractive color This theory explains color as the reflection of light by colorants or pigments, so it therefore is used in printing. It is called subtractive because as the primary colors of cyan, magenta, yellow are combined, there is a reduction in the amount of light reflected. All colors mixed together produce black (actually that's theoretical; black ink is used to compensate for imperfections in ink mixing). The absence of color allows white to reflect.

Suffix array The remaining unique parts of words after their redundant first few letters have been removed by a "front key compression" software. The idea is alphabetized lists of words can be greatly shortened if the common first few letters are removed and replaced with a code. The suffix array then becomes the critical index.

SunOS Sun Microsystems' implementation of UNIX.

Superdrive Apple's 1.44MB floppy that can read and write MS-DOS formatted floppies and Mac formatted disks.

Supered graphic A graphic superimposed over the video.

Supertwist An improved LCD display that provide a wider viewing angle and better contrast by "twisting" the crystals.

Surface modeling A CAD technique for representing solid objects.

Suspense files A workflow term. These electronic queues hold documents or files until all the required information has been gathered to process them, at which time they automatically move along to their next stop on the route.

SVGA Super Video Graphics Array. Screen resolution of 800 x 600.

S-Video Type of video signal used in Hi8 and Super VHS videotape formats. Transmits luminance and color portions separately, on different wires, thus avoiding NTSC's encoding process and its reduced quality.

Swap File An area of your hard disk that is set aside for exclusive use by Windows in 386 enhanced mode. Windows temporarily transfers information from memory to the swap file to free memory for other information. Swap files can be either temporary or permanent.

Sweep In graphic design, this tool takes a two-dimensional shape and sweeps it around an axis while scaling the shape, creating a final object. It is used to create objects such as a nautilus shell.

Swim Slow, graceful, yet undesired movements of display elements or images about their mean position on a monitor. Swim can be followed by the human eye, whereas jitters usually appears as a blur.

SWOP Specifications Web Offset Publications, a standard for representing process colors printed on offset printing presses.

Symmetrical compression Video compression that requires equal processing power for compression and decompression.

Synchronous Data transmission based on a constant time between successive bits, characters or events. The timing is achieved by the sharing of a single clock. Each end of the transmission synchronizes itself with the use of clocks and information sent along with the transmitted data. Synchronous is the most popular communications method to and from mainframes. In synchronous transmission, characters are spaced by time, not by start and stop bits. Because you don't have to add these bits, synchronous transmission of a message will take fewer bits (and therefore less time) than an asynchronous transmission. But because precise clocks and careful timing are needed in synchronous transmission, it's usually more expensive to set up synchronous transmission.

Synonym file Words grouped by their similar meanings. Allows text retrieval software to find documents in which words appear that mean the same as the query word.

Syntax The rules of grammar in any language, including computer language. Specifically, it is the set of rules for using a programming language; the grammar used in programming statements.

Synthesizer A device that produces sound from digital instructions rather than from recorded sound.

System Disk A disk that contains the MS-DOS system files necessary to start MS-DOS. You can have more than one system disk; however, the disk from which MS-DOS is loaded when you start your computer is called the boot disk.

System.INI file A Windows initialization file that contains settings you can use to customize Windows for your system's hardware.

System Partition The volume that has the hardware-specific files needed to load Windows NT. On x86-based computers, it must be a primary partition that has been marked as active for startup purposes and must be located on he disk that the computer accesses when starting up the system. There can only be one active system partition, which is denoted on the screen by an asterisk.

Moore's Imaging Dictionary

T The 20th letter of the alphabet. In ASCII, uppercase "T" is represented as hexadecimal 54; a lowercase "t" is hexadecimal 74. In EBCDIC, an uppercase "T" is hexadecimal E3; a lowercase "t" is hexadecimal A3.

Tablet Input device, used with graphics and CAD/CAE/CAM programs, that allows the user to draw using a stylus or a puck.

Tag Same as style tag. A code included in wordprocessed text that is interpreted by a desktop publishing software to control formatting, font size, style, etc.

Tagged Image File Format. See TIFF. A common bitmap file format for describing and storing color and grayscale images.

Tap A touchscreen term. A tap of the finger registers in the software as a single mouse click.

Tape A magnetic storage media. Comes in rolls or cassettes, and in standard widths: 8mm, 1/8", 1/4", 1/2". Analog music recording uses mag tape up to 2 inches wide.

Tape backup Making mag tape copies of hard disk and optical disc files, for disaster recovery purposes.

Tape drive The machine, actually a collection of devices, that transports, reads and or writes a magnetic tape.

Tape dump Printing out the raw contents of a tape storage with no attempt to format the output into reports.

Tape transport The part of the tape drive responsible for moving the tape over the read/write heads at the correct speed and tension.

TARGA Truevision Advanced Raster Graphics Adapter. An image file format.

Task List In Windows, a window that shows all the applications you have running and enables you to switch between them. You can open Task List by choosing Switch To from the Control Menu or by pressing CTRL+ESC.

TCP/IP Transmission Control Protocol/Internet Program. A set of protocols developed by the Department of Defense to link dissimilar computers across networks.

TCR Transaction Confirmation Report. A report from a fax machine listing the faxes received and transmitted. It provides details about each fax, including date, time, the remote fax's number, results, total pages.

TDMS Technical Document Management System.

Telecine The technique in which film is converted to tape. A machine that transfers film to videotape, usually using a 3:2 pulldown. The 3:2 pulldown ratio is necessary for 525/60 NTSC. For 625/50 PAL, the film is transferred straight across frame-to-frame.

Telecopier A fancy word for a facsimile machine. See fax.

Telefax 1. European term for fax. 2. A high-speed, 64 kilobit per second facsimile service that uses Group 4 fax machines and one Bearer channel of an ISDN circuit, or any other 64 kbps circuit. Group 4 fax machines take about four seconds to transmit a page. They're fast and impressive.

Tellurium A metal often used in optical discs as the recording medium.

Template A pre-designed guide to what a page should look like. Can be electronic or hard copy. Sometimes includes text or graphic elements that repeat on every page.

Terabyte From "tera," which means trillion, although it actually means 1,099,511,627,776 bytes in a computer's binary system. A terabyte is 1,024 gigabytes.

Teraflop A trillion floating point instructions per second. A measure of a computer's speed.

Moore's Imaging Dictionary

Terahertz THz. A unit denoting one trillion (10 to the 12th) hertz.

Term density Documents can be ranked in relevance based on the number of times a particular words appears, as a ratio to the total words.

Term proximity Documents can be ranked in relevance based on how close two or more words appear together.

Term weight Varying importance can be assigned to certain words, to adjust the relevance of documents in which the words appear.

Terminal emulation A setting that causes your computer to behave like a hardware terminal (a device used to display data received from a remote computer.)

Text In the context of imaging, words and numerals in electronic form.

Text-based Representation of images that requires the use of pre-existing characters rather than vector or raster graphic techniques.

Text editor An application used to create and edit simple text files. Usually works in pure ASCII, with no formatting or layout features. This dictionary was written with an ASCII text editor called ZEDIT, a derivative of the excellent QEdit.

Text file A data file consisting of alphanumeric characters, defined by a text format such as ASCII or EBCDIC. Entries in a text file are available for text searching.

Text/image retrieval The ability to locate a page image by using a full-text search. See contextual search.

Text management All the techniques and technologies involved in creating, storing and retrieving text files in an organized and logical manner.

Text search A technique for examining text files for occurrences of specific sets of characters, either in a string (a word or sentence) or in proximity (a certain word in the vicinity of another word). A "contextual search" involves finding documents based on a string of characters that appears in them. Normally, the words are linked in a relational database to an identifier for the document image from which it comes, thereby allowing image retrieval based on content.

Text-To-Speech synthesis TTS. Technologies for converting textual (ASCII) information into synthetic speech output. Used in voice processing applications requiring production of broad, unrelated and unpredictable vocabularies, e.g., products in a catalog, names and addresses, etc. This technology is appropriate when system design constraints prevent the more efficient use of speech concatenation, i.e. snippets of pre-recorded speech.

Texture mapping A two-dimensional image (such as a PICT file) is applied to the surface of an object.

Thermal wax A printing technique which uses page-sized rectangles of colored wax. A heating element in the printer melts tiny droplets of wax, which are transferred to the paper.

Thesaurus A compilation of words and phrases that compares synonymous words and terms. The basis of concept searching — if you ask for documents with "General Schwartzkopf" you can get documents where he's only referred to as "Stormin' Norman" thanks to thesauruses.

Thread A sequence of computing instructions that make up a program. A multithreaded process can have multiple threads, each executing independently and each executing on separate processors. A multithreaded program, if running on a computer with multiple processors, will run much faster than a single-threaded program running on a single processor machine. Windows NT is the first generally available multithreaded PC operating system.

Threshold A predefined level set into a scanner's software to determine whether a pixel will be represented as black or white.

Throughput The actual amount of useful and non-redundant information which is transmitted or processed. The relationship of what went in one end and what came out the other is a measure of the efficiency of that communications link — a function of cleanliness, speed, etc.

Thumbnail A small icon-like representation of an image, used to visually identify it.

TIFF Tagged Image File Format. A bitmap file format, invented by Aldus, for describing and storing color and grayscale images. Does not stand for "Took It From a Fotograph."

TIGA Texas Instruments Graphics Architecture. The TI 34010 graphics processing chip, a hardware accelerator for Windows and CAD/CAM displays (which need it). Widely used chip in commercial accelerator boards, so TIGA driver software is as close to a standard as there is in that business.

Tile In Windows, the way of arranging open windows so that no windows overlap but all windows are visible. Each window takes up a portion of the screen.

Tiling Reproducing oversize artwork or documents by breaking the image area into parts (called tiles). Adjacent tiles repeat a small portion of the image, and they may contain crop marks as well. The repeated portion of the image (the overlap) and the crop marks aid in reconstructing the overall image from the tiles.

Time code In video editing, a numbering system for video frames. The Society of Motion Picture and Television Engineers set the most common time code standard, called simply SMPTE.

Timeout If a device is not performing a task, the amount of time a computer should wait before detecting it as an error.

Timeline A horizontal line at the bottom of a video editing storyboard which is calibrated into divisions. Each division represents one frame, one second, or one minute, depending on the current scale of the timeline. You use the timeline to help position your video, audio and graphics clips on the storyboard.

Time slice The amount of processor time allocated to an application, usually measured in milliseconds. Timeslicing is how some multitasking operating systems create the impression of simultaneous processing of many tasks.

TIMS Text Information Management System. Same as full-text retrieval system.

Title bar In Windows, the horizontal bar (at the top of a window) that contains the title of the window or dialog box. On many windows, the title bar also contains the Control-menu box and Maximize and Minimize buttons.

Titles In the language of multimedia, when an author sells what he or she has created, it is called a title. The encyclopedias, dictionaries, musical works and games available on CD are all "titles." Someone authors titles. (William Safire must hate that usage!)

Moore's Imaging Dictionary

Token Ring An IBM-invented 4Mbps and 16 Mbps LAN architecture.

Toner A dry ink powder which has been electrically charged. Used in printers, fax machines and copiers. Generally, the image is translated into bit mapped charges of the opposite polarity on a special drum in the printer. The toner is attracted to the charged areas, where it is transferred to paper. The toner is then "set," usually by heat. Sometimes, the toner is attracted directly onto the paper.

Toolkit A collection of subroutine libraries that a software developer can use to create a new product. It saves him time from reinventing work someone else is willing to sell him!

Tooth The amount of roughness in a paper that allows it to absorb ink.

Touchscreen An interface that allows the user to press icons or dialog boxes with his fingertip to navigate through programs.

Touch sensitive The ability of an interface, such as a touchscreen or a digitizing tablet, to sense the pressure of the touch. Also called pressure sensitive.

Tower A PC in a vertical or upright case. Tower PCs (if they're correctly designed) have one big benefit: Heat rises and escapes more easily than in traditional horizontal machines. Heat and power surges are the most damaging threats to PC.

Track The path which is to be followed by the read head or beam during the magnetic or optical reading of a disk or disc; or the path to be followed by the recording head or beam during the writing of a disk or disc. In an optical system, the track consists of the Groove (recordable) and the Land (un-recordable).

Trackball An upside-down mouse. A rotatable ball in a housing used to position the cursor and navigate around the computer screen. A mouse needs desktop room to work, a trackball stays in one place, and can even be part of a keyboard or built into a laptop computer. It's hard to see why anyone uses a mouse instead of a trackball. This dictionary was typeset by a fine lady named Jennifer Cooper-Farrow, who used a trackball and a Macintosh computer.

Track jump The action of moving quickly from one track to another nearby.

Moore's Imaging Dictionary

Track pitch The distance between adjacent tracks' centerlines, measured radially.

Tracking 1. Adjusting the letterspacing uniformly throughout a selected portion of text. See kerning. 2. The effect created in compressed video when the speed of the transmission is not great enough to keep up with the speed of the action. Tracking creates a tearing effect on the video picture.

Tracking servo The mechanism in an optical drive which senses and adjusts for variations in movement of the recording area (the groove) of a track, caused by imperfections in the medium or the drive mechanism.

Tractor feed Continuous-form paper that has holes that are engaged by pegs or sprockets which move the paper across the platen.

Trailing A video effect in which a live video picture moves through your screen leaving a trail of still frames behind it. Very trippy.

Training In OCR, an adaptive, interactive method for recognizing printed characters. If the software does not recognize a character, it will query the user to "teach it" by manually keying in the appropriate character. Thereafter, the software will recognize that character.

Transition In video editing, how one scene or sound changes to another. Types of transitions are crossfade, cut, fade-to-black, slide and wipe.

Transmissive The way many LCD (liquid crystal display) screens on laptops reflect light.

Transparency/opacity A setting available in many image-processing functions that allows part of the underlying image to show through. 80% opacity is equivalent to 20% transparency.

Tree-structured directories A familiar name for hierarchical file management, used by both DOS and Macintosh operating systems. So-called because sub-directories can be thought of as "branching" away from the main, root directory.

Trichromatic The technical name for RGB representation of color, i.e., using read, green and blue to create all the colors in the spectrum.

Trim In video editing, to select only a portion of a clip, rather than

Moore's Imaging Dictionary

defining a new clip.

Trim in/Trim out The new beginning and ending frames on a trimmed video clip.

Trim marks Guides that show where a document will be cut to fit the specifications of a final printed product.

Trogg A Unix typesetting language.

TrueType fonts A Windows 3.1 feature. Fonts that are scalable and sometimes generated as bitmaps or soft fonts, depending on the capabilities of your printer. TrueType fonts can be sized to any height, and they print exactly as they appear on the screen. Using TrueType, you'll be able to create documents that retain their format and fonts on any Windows 3.1 machine — even if the fonts aren't installed on that computer. This makes Windows 3.1 documents portable.

TSOP Thin Small-Outline Package. A type of RAM is used in Mac PowerBook portable computers because of its compact size and low power consumption. Also, The Sound Of Philadelphia, a lush, R&B-influenced musical style invented by pop recording genius Phil Spector.

TSR Terminate and Stay Ready. A software program which can be loaded into RAM and is available for use at any time, usually triggered by a special combination of keystrokes known as a "hot key" combination. Some scanner software is TSR. AKA pop-up program.

TTS Text To Speech. A term used in voice processing. See Text-To-Speech synthesis.

TTY A teletypewriter. Typewriter-style device for communicating alphanumeric information over telecom networks.

Tumble duplex Printing both sides of a sheet so that the top of one side is at the bottom of the other.

Turbo FAT A index Netware v2.2 creates to group all the FAT (File Allocation Table) entries corresponding to a file larger than 262,144KB. The first entry in the turbo FAT index table consists of the first FAT number of the file. The second entry consists of the second FAT number of the file, etc. The turbo FAT enables a large file to be accessed quickly.

TWAIN An evolving standard for cross-platform, multi-manufacturer scanner-to-software communications. Meant to eliminate separate software drivers for every program and every scanner brand. announced in October 1991. Spearheaded by Hewlett-Packard, Logitech, Kodak, Aldus, Caere and others. Known as CLASP and "Direct Connect" during development stage. Does NOT stand for "Toolkit Without An Interesting Name." Though, maybe it should.

Twisted pair Type of cable consisting of only two wires. Used mainly for telephone lines, it is excellent and inexpensive for high-speed transmissions over short distances.

Type face Describes the design of a set of characters. Typefaces are often named for their designer: Bodoni, Goudy. Not really interchangable with font, which means the entire character set in one face and size, but many people use them interchangably without causing too much confusion.

Type font Same as font. All the characters and digits in the same style and size of type.

Type spec To name the attributes of a block of typesetting — the size, font, width, leading, alignment.

Typo The mispelling of a werd.

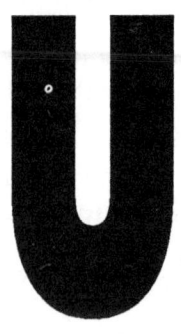

U The 21st letter of the alphabet. In ASCII, uppercase "U" is represented as hexadecimal 55; a lowercase "u" is hexadecimal 75. In EBCDIC, an uppercase "U" is hexadecimal E4; a lowercase "u" is hexadecimal A4.

UART Universal Asynchronous Receiver/Transmitter. PCs have a serial port, which is used for bringing data into and out of the computer. The serial port is used for data movement on a channel which requires that one bit be sent (or received) after another, i.e. serially. The UART is a device, usually an integrated circuit chip that performs the parallel-to-serial conversion of digital data to be transmitted and the serial-to-parallel conversion of digital data that has been transmitted.

UDK User Defined Keys.

UHF Ultra High Frequency part of the radio frequency spectrum, ranging between 300 megahertz to 3 gigahertz.

UI User Interface.

UL Underwriters Laboratories, a privately owned company that charges manufacturers a stiff fee to make sure their products meet the safety standards which UL itself develops. A UL approval means only that a product conforms to UL safety standards. It does not affirm that the product will work. UL is beginning to concern itself with adopting and promulgating standards.

Ultimedia IBM's word for the ultimate in multimedia — combining sound, motion video, photographic imagery, graphics, text and touch into a unified, natural interface representing, in IBM's words. Coined in the spring of 1992.

Ultrafiche Microfiche that can hold 1,000 document pages per sheet. Normal microfiche holds 270 pages.

Ultra hi-res Ultra high resolution. It is an imprecise term, which keeps getting higher as monitors and display controllers improve. Generally used to describe the scientific and medical imaging displays which have higher than 1280 x 1600 resolution.

UM Unscheduled Maintenance. Bad news.

Unattended Equipment working without a human operator. Many document imaging chores are repetitive and take a long time — backfile document scanning, for example. Many processing and data management chores require a dedicated machine chugging away — OCR or back up, for examples. These jobs are often best done in batches with unattended (or barely attended) devices.

Unbundled Services, programs, software and training sold separately from the hardware.

Uncoated paper Paper stock that has not been treated with a layering or calendaring process to give it a glossy sheen. Used by laser printers and copiers that require more "tooth" in the paper to accept toner.

Underlap In facsimile, a defect that occurs when the width of the scanning line is less than the scanning pitch.

Underscan The part of an image that falls inside the borders of the display screen, i.e., the part you can see.

Underscore Typesetting with a horizontal line under it. Used for emphasis. Complicates OCR.

Underset Text that falls too short. Given a choice, I'll take underset over overset any day of the week.

Undo buffer The portion of memory that holds a version of a document before the most recent changes were made. Used to restore the previous version if — when — you make a mistake.

Unerase An MS-DOS command for getting back files you've accidentally erased.

Unfragmented A hard disk that has most of its files stores in consecutive sectors, rather than spread out all over the disk in an interleaved fashion.

Unit spacing Same as fixed spacing.

Unix A general-purpose, multi-user, multitasking operating system invented in 1969 by Ken Thompson of AT&T Bell Labs. Unix is powerful and complex, and needs a computer with a large amount of RAM memory to support its power. Unix allows a computer to handle multiple users and multiple programs simultaneously. And it works on many different computers, which means you can often take applications software and move it — with little changing — to a bigger, different computer, or to a smaller, different computer. This process of moving programs to other computers is known as "porting." Today, Unix is available on a wide range of hardware, from small personal computers to the most powerful mainframes, from a multitude of hardware and software vendors. Most people spell UNIX all caps. I don't, because it's not an acronym for anything.

Unjustified Typesetting in which neither side aligns, i.e, both sides are ragged.

Unstructured data Data variable length that is subject to change at any moment. Text is unstructured data. To contrast, fixed-length fields in a database is structured data.

Unzip To decompress a file using the popular compression software called PKUNZIP.

UPC Universal Product Code. A standard machine-readable bar code, applied to retail products for inventory and pricing management purposes.

Upper memory area The 384K area of address space adjacent to the 640K of conventional memory. This area is usually reserved for running your system's hardware, such as your monitor, and is not considered part of total memory because applications cannot store information in this area. In 386 enhanced mode, Windows can access unused portions of this area.

Upper memory blocks Areas of the upper memory area that contain general-purpose memory and that can be used to hold device drivers or other memory-resident programs in order to leave more conventional memory available for applications.

V The 22nd letter of the alphabet. In ASCII, uppercase "V" is represented as hexadecimal 56; a lowercase "v" is hexadecimal 76. In EBCDIC, an uppercase "V" is hexadecimal E5; a lowercase "v" is hexadecimal A5.

V Abbreviation for volt.

VAB Value Added Business partner. A term which Hewlett-Packard uses for developers who write software for its computers. HP helps its VABs sell software. Clearly, by doing so, it helps sell more HP computers.

Value The relative lightness or darkness of a color.

Value driven re-engineering A fancy term for re-engineering, which is a term probably invented by Michael Hammer in the July-August, 1990 issue of Harvard Business Review. The term now seems to mean taking tasks presently running on mainframes and making them run on file servers running on LANs — Local Area Networks. The idea is to save money on hardware and make the information more freely available to more people. More intelligent companies also redesign their organization to use freely available information.

Vaporware A semi-affectionate slang term for software which has been announced and perhaps even demonstrated, but not delivered to commercial customers.

Variable field A database field that can accept an unlimited number of characters. Used in "memo" or "comment" fields.

Variable length record A file in a database containing records not of uniform length and in which the distinctions between fields are made with commas, tabs or spaces (called "delimited.") Records become uniform in length either because they are uniform to start with or they are "padded" with special characters.

Variable text A field of text which changes throughout the process of a job. The names and address in a form-letter printing job, which changes each time a new letter prints, is an example.

VAR/VAD Value Added Reseller/Dealer. VARs buy equipment from computer manufacturers, add some of their own software and possibly some peripheral hardware to it, then resell the system, with its newly added "value" to end users. A VAD (Value Added Dealer) is similar, but is generally less directly in touch with the end user.

VCR Video Cassette Recorder. A magnetic recording and playback machine. Generally used for recording and viewing full-motion video, but useful (when adapted) as a data backup device.

VDISK Virtual DISK. Part of the computer's Random Access Memory assigned to simulate a disk. VDISK is a feature of the MS-DOS operating system.

VDT Video Display Terminal. Generic, slightly inaccurate, name for any display terminal.

VDU Visual Display Unit. Another term for a computer monitor. VDU is preferred in Europe.

Velocity of light The speed of light in a vacuum is 186,280 miles per second, or 299,792 kilometers per second. The speed of light is very important because today we can measure time more accurately than length. In effect, we define the meter as the time traveled by light in 0.000000003335640952 of a second as measured by the cesium clock.

Velocity of sound The velocity of sound varies with the medium carrying it. In air at 0 degrees centigrade, it's 331 meters per second. In glass at 20 degrees centigrade, it's 5,485 meters per second.

VTR Video Tape Recorder.

Vector Images defined by sets of straight lines, defined by the locations of the end points. Used in CAD and engineering software for precise, mathematically editable designs.

Moore's Imaging Dictionary

Vector display Terminal that displays images with vectored line segments, rather than pixels.

Vectorization Translation of a pixel-based image to a vector-based image, usually to be compatible with a CAD program.

Velox A halftone print. Used to paste-up mechanicals with halftones in place, saving negative stripping work, and costs. Downside: the quality is lower.

Ventura Publisher A popular desktop publishing program for PCs from Ventura Software, Inc., a division of Xerox. Outputs PostScript. Best known for its ability to paginate large documents. See Quark Xpress.

Verso A left-hand page. Always even numbered.

Vertical blanking interval The interval between television frames in which the picture is blanked to enable the trace (which "paints" the screen) to return to the upper left hand corner of the screen. Several companies are eyeing the vertical blanking interval as a place to send digital data, including news and weather information.

Vertical recording A magnetic disk recording technique that increases the available storage space.

Vertical redundancy check VRC. A check or parity bit added to each character in a message such that the number of bits in each character, including the parity bit, is odd (odd parity), or even (even parity). See PARITY.

Vertical scan frequency Same as refresh rate, expressed in hertz.

Vertical tab Spacing down a preset number of lines to a new location on the page. Like regular tabbing, except down instead of across.

VESA Video Electronics Standards Association. Along with eight leading video board manufacturers, NEC Home Electronics founded VESA in the late 1980s. The association's main goal is to standardize the electrical, timing and programming issues surrounding 800 x 600 pixel resolution video displays, commonly known as Super VGA. VESA has also issued a standard called "local bus," a high-speed bus for the PC designed to move video between the CPU and the screen a lot faster than the conventional AT bus.

Vesicular film Similar to diazo film (it uses diazo salts instead of silver), vesicular film is used to make duplicates of microfilm which are the opposite polarity of the original.

VGA Video Graphics Array. Standard IBM video display standard. Screen resolution is 640 x 480. Provides medium-resolution text and graphics. VGA is superior to earlier graphics standards, such as CGA and EGA. VGA is barely adequate for imaging.

VHD Very High Density. Techniques of recording 20 megabytes and more on a 3 1/2" magnetic disk.

VHS Very High Frequency. Frequencies from 30 MHz to 300 MHz.

Video camera Camera capable of acquiring and deliver to a recording device full-motion video. Converts the moving image into a series of horizontal lines, which are further broken down into continuously varying pixels for display. Most often analog, but digital video cameras which use charge coupled devices (CCDs) are becoming common.

Video capture Converting a video signal into a format that can be saved onto a hard disk or optical storage device and manipulated with graphics software. This is accomplished with a device internal in a computer called a "frame grabber" or video capture board. Images thus captured are digitized, and can be treated as any digital object.

Video codec The device that converts an analog video signal into digital code.

Video conferencing The real-time, and usually two-way, transmission of digitized video images between two or more locations. Transmitted images may be freeze-frame (where television screen is repainted every few seconds to every 20 seconds) or full motion. Bandwidth requirements for two-way videoconferencing range from 6 MHz for analog, full-motion, full-color, commercial grade TV to two 56 Kbps lines for digitally encoded reasonably full motion in full color, to 384 Kbps for even better video transmission to 1,544 Mbit/s for very good quality, full-color, full motion TV.

Video digitizer Same as frame grabber; A device that changes a video picture into a digital computer graphics language.

Videodisc Read-only, 12" direct access optical disc standard for storing and playing video.

Moore's Imaging Dictionary

Videotex Two-way interactive electronic data transmission or home information retrieval system using the telephone network. Videotex has not been successful because of its (erstwhile) need for expensive, proprietary (i.e. dedicated) equipment and lack of variety in information offered.

Video Toaster A remarkable software/hardware product from NewTek, Topeka, KS, that allows you to edit and add powerful special effects to video using a Commodore Amiga computer. An example of the "democratization" of technology: broadcast- and film-quality special effects are possible with a Toaster that required million-dollar production facilities only a couple years ago. Called the Toaster because the makers want to position it as a common "appliance."

VIM The Vendor-Independent Messaging Group. VIM includes Apple, Borland, IBM, Lotus, MCI Mail, Novell and WordPerfect. VIM's collaborates on developing an open, industry-standard interface that will allow e-mail features to be built into a variety of software products.

VINES Virtual networking software which is the core of the LAN from Banyan, Westboro, MA.

VIP 1. Vector Image Processor. 2. Ventura Integrated Picture, a feature in their PicturePro product that allows you to keep "draft" version of images until you're ready to finally render them. It does not stand for "Verks In Progress."

Virtual drive A portion of RAM used as if it were a hard disk drive. Virtual drives are much faster than hard disks because your computer can read information faster from memory than from a hard disk. Information on a virtual drive is lost when you turn off or restart your computer. Also known as RAM drive.

Virtual memory A memory-management system used by Windows in 386 enhanced mode that enables Windows to run as if there were more memory than is actually present. The amount of virtual memory available equals the amount of free random-access memory (RAM) plus the amount of disk space allocated to a swap file that Windows uses to simulate additional RAM.

The benefit of using virtual memory is that you can run more applications at one time than your system's physical memory would otherwise allow. The drawbacks are the disk space required for the virtual-memory swap file and the decreased execution speed because of the swapping.

Virtual reality The publisher of Virtual Reality Report says, "Virtual reality is a way of enabling people to participate directly in real-time, 3-D environments generated by computers." Virtual reality involves the user's immersion in and interaction with a graphic screen/s. Using 3-D goggles and sensor-laden gloves, people "enter" computer-generated environments and interact with the images displayed there. Once virtual reality was called artificial reality. But artificial means "fake," while virtual means "almost."

Virtual storage Storage space that may be viewed as addressable main storage to a computer user, but is actually auxiliary storage (usually peripheral mass storage) mapped into real addresses. The amount of virtual storage is limited by the addressing scheme of the computer.

Virus A software program capable of replicating itself and usually capable of wreaking great harm on the system.

VLF Very Low Frequency. That portion of the electromagnetic spectrum having continuous frequencies ranging from about 3 Hz to 30 kHz.

VLSI Very Large Scale Integration. Semiconductor chip with several thousand active elements or logic gates — the equivalent of several thousand transistors on a single chip. VLSI is the technique for making the so-called "computer on a chip."

VM Voice Mail. Voice Messaging. Virtual Memory.

VME A bus commonly found on mainframe computers.

Volatile storage Computer storage that is erased when power is turned off. RAM is volatile storage.

Volt The unit of measurement of electromotive force. Voltage is always expressed as the potential difference in available energy between two points. One volt is the force required to produce a current of one ampere through a resistance or impedance of one ohm.

Volume A partition or collection of partitions that have been formatted for use by a computer system. A Windows NT volume can be assigned a drive letter and used to organize directories and files. In

NetWare a volume is a physical amount of hard disk storage space. Its size is specified during installation. NetWare v2.2 volumes, for example, are limited to 255MB and one hard disk, but one hard disk can contain several volumes. A NetWare volume is the highest level in the NetWare directory structure (on the same level as a DOS root directory). A NetWare file server supports up to 32 volumes. NetWare volumes can be subdivided into directories by network supervisors or by users who have been assigned the appropriate rights.

Volume label A name assigned to a floppy or hard disk in MS-DOS. The name can be up to 11 characters in length. You assign a label when you format a disk or, at a later time, using the LABEL command.

VRAM Video RAM. Special memory used to support high resolution, fast refresh video.

VSA Virtual Storage Access Method.

VTAM Virtual Telecommunications Access Method. (Pronounced "VEE-tam.") A program component in an IBM computer which handles some of the communications processing tasks for an application program. VTAM also provides resource sharing, a technique for efficiently using a network to reduce transmission costs.

VTR Video Tape Recorder.

VU meter Volume Unit meter measures the strength of an audio signal. Old-fashioned ones had a needle which moved from left to right, like the "applause meter" on American Bandstand. New ones have chasing LEDs.

Moore's Imaging Dictionary

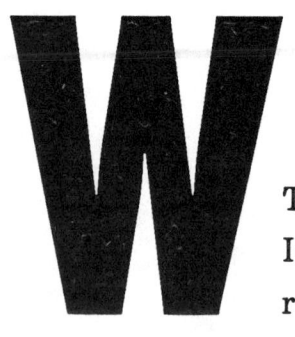

W

The 23rd letter of the alphabet. In ASCII, uppercase "W" is represented as hexadecimal 57; a lowercase "w" is hexadecimal 77. In EBCDIC, an uppercase "W" is hexadecimal E6; a lowercase "w" is hexadecimal A6.

W Abbreviation for watt.

Wallpaper In Windows, the pattern or image displayed on the desktop background.

WAN Wide Area Network. Uses common carrier-provided lines that cover an extended geographical area. Contrast with LAN. This network uses links provided by local telephone companies and connects disperse sites.

Wand A hand-held scanner used for OCR or for reading bar codes.

Warm When referring to color, it means there is too much red in the image.

Wave A pattern inherent in certain types of energy, including light and sound. Waves have several characteristics, including amplitude ("height" of the wave), wavelength ("width" of the wave) and frequency (number of times it cycles in a given period of time). Frequency is measured in hertz, or cycles per second.

Watt The unit of electrical power and representing the product of amperage and voltage.

Waveform The visual representation of an analog signal, showing its amplitude (height) and its frequency (width).

Waveform monitor A specially designed oscilloscope display that graphically shows waveforms.

Web offset A widely used means of printing large-run books, magazines and documents. Uses large rolls of paper (called webs) which move through the printing press or presses (in the case of multiple colors). The presses are of the offset lithography type, meaning they print by transferring the ink from a plate to a rubber sheet, then to the paper (hence, the image is "offset").

Weight In typography, the relative thickness or blackness of the characters.

Width table The part of the publishing software that keeps a listing of the widths of a font's characters; used for proportional spacing.

Widow A word, partial word or letter that falls by itself on the last line of a paragraph. Typographers and graphic artists love to hunt out widows and "orphans."

Wildcard A character in a text search that stands for other characters. For instance, a search for GEO* (with the asterisk being the wildcard) would find all occurrences of words starting with the letters GEO — geography, geostationary, geology, etc.

WIMP interface Stands for Windows, Icons, Menus and a Pointing device. A derogatory reference to GUI.

Winchester disk The common version of the sealed hard disk. Invented by IBM. Winchester was the code name during its development, because the name of the system it was designed for (3030) reminded somebody of the Winchester rifle.

A Winchester hard disk drive consists of several "platters" of metal stacked on top of each other. Each of the platter surfaces is coated with magnetic material and is read and written to by heads which float across (but don't touch) the surface.

Window A rectangular area on your screen in which you view an application or document. You can open, close and move windows. You can open several windows at a time, and you can often reduce a window to an icon or enlarge it to fill the entire desktop. Sometimes windows are displayed within other windows.

Windows application An application that runs under Windows and does not run without Windows. All Windows applications follow similar conventions for arrangement of menus, style of dialog boxes and keyboard and mouse use.

Windows A Microsoft operating "environment" (not an operating system) that features multiple screens and a graphical user interface (GUI). Led to much user-pleasing software development, because of its consistent familiar interface and emphasis on mouse-driven point-and-shoot operation.

Windows NT A true network operating system, introduced by Microsoft in the summer of 1993, characterized by 32-bit architecture, network features and multi-tasking.

Windowing A technique for reducing data processing requirements by electronically defining only a small portion of an image for analysis, ignoring the rest.

WIN.INI file A Windows initialization file that contains settings you can use to customize your Windows environment.

Wipes Transition effects in video, animation or presentation programs. Wipes are methods of ending one scene and moving to another. They clear the screen in one of several ways, including fade-outs, flashes and sweeps.

Wire printer A matrix printer which uses a set of wire hammers to strike the page through a carbon ribbon, generating the matrix characters.

Wireframe modeling A CAD and 3D design technique. The first step in a design. The shape of the object is developed with many intersecting lines (it looks like it's made from chicken wire.) Later, surface colors and textures (called "shaders") are applied to the wireframe models for a more realistic look.

Word A collection of bits the computer recognizes as a basic information unit and uses in its operation. Usually defined by the number of bits contained in it, e.g. 8, 16 or 32 bits. Using DOS, the IBM PC defines a word as eight bytes.

Word spotting In speech recognition, the process whereby specific words are recognized under specific speaking conditions (i.e. natural, unconstrained speech). Can refer to the ability to ignore extraneous sounds during continuous word recognition.

Word wrap A feature that moves text from the end of a line to the beginning of a new line as you type. With word wrap, you do not need to press ENTER at the end of each line in a paragraph.

Wordprocessing Software that controls the input of text and the formatting of its output. I like to spell it as one word, no space. i do that because it's a category of technology. I think the activity — "She does word processing as a night job" — should be two words.

Workflow Automating the orderly procedures for handling business processes. Worklow systems are usually based on electronic versions of documents — how they are routed through departments in a company; which transactions have to be accomplished in which order, what to do about exceptions and mistakes — are all workflow concerns. Workflow software schedules processing, routes documents automatically among departments and tracks document status.

Workprint In video editing, a second generation copy of the original, used to avoid wear-and-tear on the original during rough editing.

Workspace Windows' word for the area that displays the information contained in the application or document you are working with.

Workstation A single-user microcomputer or terminal, usually one that is dedicated to a single type of task (graphics, CAD, scientific applications, etc.).

WORM Write Once Read Many. Optical storage technique in which data is permanently recorded. Data can be erased, but not altered.

Write head A magnetic head capable of writing only. You find write heads on everything from tape recorders to computers.

Write protect Using various hardware and software techniques to prohibit the computer from recording (writing) on storage medium, like a floppy or hard disk. You can write protect a 5 1/4" diskette by simply covering the little notch with a small metal tag (physical write protect). You write protect a hard disk file with software (logical write protect).

Wrong reading Reversed, right-to-left image reproduction. Negatives are often wrong reading. Contrast with right reading.

WYPIWYF Acronym for "What You Print Is What You Fax," also "The Way You Print Is the Way You Fax." Coined by Intel to describe its one-step pop-up menu that makes sending faxes from the PC as easy as sending a document to a printer.

WYSIWYG What You See Is What You Get. One of the acronyms the computer industry is famous for. Pronounced "wizzy-wig," it refers to a graphics or publishing program that displays images on the screen (nearly) exactly the way they will appear on paper. True WYSIWYG is becoming more of a reality in 1993, thanks to color-calibrating monitors, and monitors which display the CMYK (print) color model.

Y The 25th letter of the alphabet. In ASCII, uppercase "Y" is represented as hexadecimal 59; a lowercase "y" is hexadecimal 79. In EBCDIC, an uppercase "Y" is hexadecimal E8; a lowercase "y" is hexadecimal A8.

Y/Alpha delay In video editing, an adjustment that synchronizes the luminance channel with the alpha channel, to compensate for processing delays caused by encoding.

Y-dimension of recorded spot In facsimile, the center-to-center distance between two recorded spots measured perpendicular to the recorded line.

Y-dimension of scanning spot In facsimile, the center-to-center distance between two scanning spots measured perpendicular to the scanning line on the subject copy.

Yellow One of the colored inks used in four color printing. One of the subtractive process colors; reflects red and green and absorbs blue.

Yellow Book Another name for the ISO 9660 standard for CD-ROM creation, set in 1987. It described how a table of contents should be organized, but not the actual file format of the data. Has led to incompatibilities among CD-ROM vendors. Also called the "High Sierra" standard.

Ymodem A faster transfer variation of Xmodem. Instead of a 128-byte block, Ymodem sends 1,024 bytes (1 kilobyte). Ymodem combines the 1K block and the 128-byte block modems into the same protocols. Ymodem, or 1K as it is known, is a thrifty way to send files over the telephone.

YUVA A color model for video that describes the luminance (Y) two color channels (U and V) and the opacity (A).

Z The 26th letter of the alphabet. In ASCII, uppercase "Z" is represented as hexadecimal 5A; a lowercase "z" is hexadecimal 7A. In EBCDIC, an uppercase "Z" is hexadecimal E9; a lowercase "z" is hexadecimal A9.

Zap Informal for wipe out, as in to zap a file by accidentally overwriting it.

Zapf dingbats A popular dingbat, or ornament, font. Some Zapf dingbats are: ● ☛ ✓ ✪ ✚ ✸ ✂ ✝ ✎

Zero The numeral is often written with a slash through it to to distinguish it from the uppercase letter "O."

Zero bit The high-order bit in a byte or a word.

Zerofill Here's a definition from GammaLink, a fax board maker: A traditional fax device is mechanical. It must reset its printer and advance the pages as it prints each scan line it receives. If the receiving machine's printing capability is slower than the transmitting machine's data sending capability, the transmitting machine adds "fill bits" to pad out the span of send time, giving the slower machine the additional time it needs to reset prior to receiving the next scan line.

Zero slot lan A Local Area Network (LAN) that uses a PC's serial port to transmit and receive data. It doesn't require a network interface card to be installed in a slot in the PC, thus the name "zero-slot" LAN. RS-232 LANs usually use standard RS-232 or phone cable to link PCs. Software does the rest of the work. Due to the slow speed of serial communications on a PC, RS-232 LANs are usually restricted to speeds of around 19.2K bits per second. What they lose in speed, however, they make up in low price.

Zip To compress a file using the popular program PKZIP.

Z line The hypothetical line on which characters rest — the baseline.

Zmodem A fast error-correcting file transfer/data transmission protocol for transmitting files between PCs. Zmodem does full duplex-transmission. It does not depend on any acknowledgement (ACK) signals from the host computer.

Zoom in To enlarge a portion of an image in order to see it more clearly or make it easier to alter. You see less of the overall image.

Zoom out gives you a global overall view of an image, but less detail.

Moore's Imaging Dictionary

Numbers

Prefixes:

Atto = 1×10^{-18}
Femto = 1×10^{-15}
Pico = 1×10^{-12}
Nano = 1×10^{-9}
Micro = 1×10^{-6}
Milli = 1×10^{-3}
Centi = 1×10^{-2}
Deci = 1×10^{-1}
Deca = 1×10^{1}
Hecto = 1×10^{2}
Kilo = 1×10^{3}
Mega = 1×10^{6}
Giga = 1×10^{9}
Tera = 1×10^{12}

1.2M The 1.2 megabyte high-density 5 1/4" floppy disks.

1.4M The 1.4 megabyte high-density 3 1/2" disk cartridges, or "microfloppy."

1.544 MBPS The speed of a T-1 telecommunications circuit.

2B+D A shortened way of saying ISDN's BRI interface, namely two bearer channels and one data channel. See ISDN.

3 1/2" The diameter of a floppy disk, also called a common disk cartridge, or "microfloppy."

4:2:2 A digital video standard where the luminance is sampled at 720 times a line and the two chrominance channels (R-Y and B-Y) are sampled at 360 times a line. The "4" and "2" relates to the multiple frequencies of the color subcarrier signal. A system for sampling the frequencies used to digitize the luminance and color difference components of a component video signal.

4:4:4:4 A digital video standard where the luminance both chrominance channels, and a key or alpha channel are all sampled at full bandwidth. The symbol 4:4:4:4 signifies that the analog-to-digital sampling, digital filtering, all graphics operations, image manipulation and digital storage are performed on all four channels (Y,R-Y,B-Y,A) of the video signal. This sampling is done at the rate of 13.5 MHz. Benefits are: superior image expansion quality, superior hard- and soft-edged key quality and more accurate treatment of colors when transforming images.

4FSC The frequency used for sampling when digitizing a composite signal (4 times subcarrier frequency). For NTSC this sampling frequency is 14.3 MHz, for PAL this frequency is 17.7 MHz.

4-field sequence Refers to the property of NTSC composite video to require four successive fields to repeat a field with identical subcarrier phase-to-sync relationship. Insert or assemble editing on a field that is not the correct field in the sequence will product a horizontal picture shift at the edit frame.

4GL Fourth Generation Language. A programming language that commands the computer at a higher level abstraction than normal high-level languages.

5 1/4" The diameter of a common floppy disk.

7-bit ASCII The standard code for text in which a byte (eight bits) holds the seven ASCII digits that define the character plus one bit for parity.

8-bit grayscale To understand, think first about one bit. One bit can tell you just one of two possible things about a pixel: either it's on (black) or it's off (white). If you assign 8 bits to describe that pixel, it can tell you 256 things about it — that's the number of possible combinations of eight bits being either on or off. If you assign values, expressed in percentages of gray, to each bit, those 256 combinations can stand for 256 shades of gray — actually, black (0), white (256) and 254 steps in between.

8-bit color Also called indexed, or adaptive color. Software analyzes a (usually 24-bit) scan, and chooses the 256 most commonly appearing colors. It is a color image compression technique.

9-track A standard for 1/2" magnetic tape designed for data storage. Its nine tracks hold a byte (eight bits) plus one parity bit in a row across the tape width-wise.

10 Base-T An Ethernet local area network which works on twisted pair wiring that looks and feels remarkably like telephone cabling. Sometimes old phone cabling will work. Mostly, it won't. Be safe. Put in new cabling.

10 bits Quantizing levels allowed by some digital video standards. 10Bit quantizing results in 1024 levels of grayscale video.

19MM cassette The standard cassette used for both D1 and D2 digital videotapes (the tape is 19mm wide). D1 and D2 cassettes are not compatible with each other.

23B+D A circuit with a wide range of frequencies that is divided in 23 64 Kbps paths for carrying voice, data video or other information simultaneously. In ISDN, it is also known as the Primary Rate Interface.

24-bit color First read the definition of 8-bit grayscale. Now imagine if each pixel on the screen could be assigned 24 bits — 8 for each additive color red, green and blue. That would mean there are 256 x 256 x 256 possible combinations, or 16,777,216 possible colors each pixel could be. Remember: it's 16.8 or so million POSSIBLE colors...a 24-bit scan may or may not (probably does not) ACTUALLY have that many shades in it.

24-bit mode The standard addressing mode of Apple Macintosh's System 6 operating system, where only 24 bits are used to designate addresses. Limits address space to 16MB (2 to the 24th power), of which only 8MB is normally available for application memory. This mode is also used under System 7 (the Mac's present operating system) if 32-bit addressing is turned off. See 32-BIT.

24 to 8 Slang for the act of indexing color as a compression technique. A 24-bit color scanned image has 16.8 million or so possible colors. Software analyzes the image and extracts the 256 (8 bit) most commonly appearing colors. Think of all 16.8 million color possibilities being placed on a bell curve. Find the 256 most common colors in the middle of the curve. Now chop off the ends. That's 24 to 8.

Moore's Imaging Dictionary

32-bit An adjective that describes hardware or software that manages data, program code and program address information in 32-bit-wide words.

32-bit addressing An optional addressing mode available in Apple Macintosh System 7 operating system (or System 6 if OPTIMA is installed). Because 32 bits are used to designate address information, up to 256MB of physical RAM or up to 1,024MB (one gigabyte) of virtual memory on the Quadra 900 can be addressed. Also called 32-bit mode.

32-bit color Along with the 24 bits of color (8 for red, 8 for green and 8 for blue) another 8 bits is used to describe the opacity (solid) or transparency (clear) or 254 levels of in between. Used by new high-end photo editing and painting packages to assign composite images different layers, for easier manipulation and special effects.

32-megabyte barrier Versions of MS-DOS prior to 4.0 had a built-in limit on the size of a disk partition. The original design parameters of DOS used one 16 bit "word" to access sectors within its hard disk partition. A 16-bit word can represent values from zero through 65,535. This limits the partition's total sector count to 65,536. Hard disk sectors are 512 bytes long. 512 bytes times 65,536 sectors = 33,554,432. Since there are 1,048,576 bytes in each megabyte, the maximum size partition calculates to 32 megabytes. MS-DOS can now create and read partitions of up to 512 megabytes.

119 Japan's equivalent of the United States' emergency 911 number.

286 Short for Intel's 80286 microprocessor chip. Unveiled in February, 1982, the 80286 was the second generation (thus the 2) 8086 chip. The 286 (sometimes written '286) featured 134,000 transistors. It is a 16-bit processor. This chip will run at 6 MHz to 16 MHz. At 10 MHz it runs 1.5 mips. The 80286 has a 24-bit address bus and can address 16 MB of memory directly (which is 2 raised to the 24th power).

360K The standard capacity of a regular density 5 1/4" floppy disk, now effectively obsoleted by 1.2 meg "high density" floppy disks.

386 The shortened name for Intel's 80386 microprocessor. Introduced in October, 1985, the 386 (sometimes written '386) is the third generation of the 8086 family (hence the 3). It contains 275,000 transistors and is a 32-bit processor. It is typically clocked at 16 MHz, but there are now versions of it that will run at 33 MHz. Its 32-bit processor means is it can process data in 32-bit chunks, rather than the 16-bit chunks of the 80286 family. The 386's main innovation is its memory management ability, which improves performance by enabling the CPU to work with memory segments larger than 64K and to have access to more than one megabyte of memory (both limitations of the 286). The 386 can address up to four gigabytes of physical memory and 64 terabytes of virtual memory at a time.

386 enhanced mode A mode in which Windows appears to use more memory than is physically available and provides multitasking for non-Windows applications.

486 A shortened name for Intel's family of 80486 chips, the successor to and continuation of the line of chips that started with the 8088 and grew into the 8086, the 80286 and the 80386. In simple terms, the 486 is a combination microprocessor, floating point math coprocessor, memory cache controller and 8K of RAM cache, all in one chip. The 486 chip contains 1.2 million transistors and is capable of 41 MIPS operating at 50 MHz.

486DX The full-powered 80486 from Intel.

486SL A power-saving version of Intel's '486 chip, which it introduced in the fall of 1992 and targeted at manufacturers of laptops and notebooks.

486SLC A name given to clones of 80486 chips. These clones are made by companies other than Intel, the manufacturer of the original chip. These clones are often not identical in power with the Intel chip.

486SX The Intel 486SX processor is essentially a full-powered 486DX with its math coprocessor disconnected. Unlike the 386 line, in which the SX version used a 16-bit external path, Intel created the 486SX by simply disconnecting the 486DX's math coprocessor. The 486SX shares the DX's full 32-bit data path and 8K cache.

720K The standard capacity of a 3 1/2" disk or disk cartridge, or "microfloppy," now obsoleted by the 1.44 megabyte "high density" floppy disks.

999 Great Britain's equivalent of the United States emergency number 911.

247

Moore's Imaging Dictionary

8080 Intel brought out the 8-bit 8080 microprocessor in April, 1974. It contained 5,000 transistors and was 10 times more powerful than the 8008. The 8080 was, probably, the "first" microprocessor.

8086 On June 8, 1978, Intel introduced the 8086, a 16-bit microprocessor with 29,000 transistors and 10 times the performance of the 8080. The 8086, or more precisely its 16-bit counterpart, the 8088, became the brain of IBM's first personal computer.

8088 Introduced in June, 1979, by Intel, this is the 16-bit chip which drove the original IBM PC and the subsequent IBM XT. It is identical to the 8080 except that external bus width is 8-bit (the 8080 is 16). The 8088 has a clock speed of 4.77 MHz at 0.33 MIPS and 8 MHz at 0.75 MIPS.

68020 A Motorola microprocessor used in the Macintosh II and LC. On the Mac II there is a socket to hold a coprocessor called a PMMU which enables the Mac II to use virtual memory. There is no socket on the LC motherboard, so the processor must be upgraded in order to use virtual memory.

68030 A Motorola microprocessor used in the SE/30, PowerBook 140/145 and 170, and Modular Macs, except the LC, Mac II, and Quadras. It supports virtual memory without any additional hardware.

68040 A Motorola microprocessor used in Quadras. It supports virtual memory without any additional hardware. Has built-in floating point math capabilities (an optional coprocessor on the 68030).

80486SX The Intel 80486 microprocessor without a math coprocessor. The 80486SX is a cheaper version of the 80486 chip.

80586 An Intel microprocessor containing more than three million transistors, introduced in April, 1993 and called the Pentium. It is 80% faster than the fastest 486. It is capable of executing 112 million instructions per second.

80686 An upcoming Intel microprocessor containing more than seven million transistors. It will be capable of executing 175 million instructions per second. It is due out in 1993/1994.

80786 An upcoming Intel microprocessor containing more than 20 million transistors. It will be capable of executing 250 million instructions per second. It is due out in 1995/1996.

MOORE'S IMAGING DICTIONARY

Dates

1793 Semaphore invented.

1840 Samuel Morse patents the telegraph.

1843 First successful fax machine patented by Sottish inventor, Alexander Bain. His "Recording Telegraph" worked over a telegraph line, using electromagnetically controlled pendulums for both a driving mechanism and timing. At the sending end, a stylus swept across a block of metal type, providing a voltage to be applied to a similar stylus at the receiving end, reproducing an arc of the image on a block holding a paper saturated with electrolytic solution which discolored when an electric current was applied through it. The blocks at both ends were lowered a fraction of an inch after each pendulum sweep until the image was completed.

1844 Samuel Morse send sends first public telegraph message.

1865 First commercial fax service started by Giovanni Casselli, using his "Pantelegraph" machine, with a circuit between Paris and Lyon, which was later extended to other cities.

1927 April 7, 1929. First public demonstration of video phone technology. Moving black and white pictures were sent over telephone wires between Secretary of Commerce Herbert Hoover in Washington DC and AT&T executives in New York. They went at 18 frames per second. Further development of this technology led to the creation of TV.

1947 The year the transistor was invented.

1960 The year the laser was invented. By Theodore Maiman of the U.S. Laser stands for Light Amplification by the Stimulated Emission of Radiation.

1964 Prototype of the first video phone made by the Bell System shown at The World's Fair in Queens, New York City. Pictures were black and white and the technology was very expensive.

1966 October, 1966 the Electronic Industries Association issues its first fax standard: the EIA Standard RS-328, Message Facsimile Equipment for Operation on Switched Voice Facilities Using Data

Communications Equipment. The Group 1 standard, as it later became known, made possible the more generalized business use of fax. Transmission was analog and it took four to six minutes to send a page.

1975 Bell System begins testmarketing Picturephone, a two-way color videoconferencing service at 12 locations around the country. Businesses rented meeting rooms equipped with the technology.

1977 First lightwave system installed.

1978 CCITT comes out with Group 2 recommendation on fax.

1980 CCITT comes out with Group 3 recommendation on fax. Group 3 machines are much faster than Group 2 or 1. With Group 3 machines, after an initial 15-second handshake that is not repeated, they can send an average page of text in 30 seconds or less.

1981 The year the IBM PC debuted.

1983 Novell's first network file service software. Nintendo introduces Famicom, a computer-turned-video game.

1986 Novell's SFT NetWare, first fault tolerant local area network operating system.

1989 Novell releases NetWare 3.0, the first 32-bit network operating system for Intel 80386/486-based servers.

1992 1. AT&T introduces VideoPhone 2500 marketed as the first home-model color video phone which works on normal dial up analog phone lines. It meets cool reception because of poor image quality and its high price, namely $1,500.

2. Microsoft Windows 3.1 and IBM's OS/2 2.0 operating systems introduced. Windows NT (32-bit operating system) debuts in beta form. 3. Wang files for Chapter 11. 4. MCI introduces VideoPhone for normal dial-up analog phone lines. It retails for $750. It is not compatible with the AT&T phone.

1993 Microsoft debuts the Windows NT operating system.

Moore's Imaging Dictionary

Marks

The character usually in the "shift 3" position on a QWERTY keyboard. It's commonly called the pound sign, but it's also the "number" sign, the "crosshatch" sign, the "octothorpe" sign and the "tic-tac-toe" sign.

& The ampersand should never be used to mean "plus" ("+"). An ampersand joins two elements (Sonny & Cher); a "plus" sign accumulates values (2 + 2).

* An asterisk is often used to represent a wild card. For example, the command

    ERASE JOHN.*

will erase all the files on your disk beginning with JOHN, e.g. JOHN.TXT, JOHN.NEW, JOHN.OLD, JOHN.BAK, etc.

- The hyphen is not the same as a dash ("—"). The hyphen joins two adjectives combined to describe one noun: "A magneto-optical disc." Technology writing creates a lot of new words by dropping the hyphen as familiarity grows. One day we may shift to "magnetooptical." Maybe that was a bad example....

Also used as a "minus" sign, or to signify negative value of a number.

— The dash is used to signify a range between two numbers (12 — 50). It's also used — by me, especially — to set off an appositive, or additional thought.

/ The forward slash. Lotus made it famous.

@ The character usually in the "shift 2" position on a QWERTY keyboard. It's called the "at sign." Its biggest use these days is in spreadsheets and in desktop publishing.

**** The backslash. Used for designating directories on your MS-DOS machine. This dictionary is located in

    C:\WORK\DICTIONA>

That means it's in the "dictiona" subdirectory of the "work" directory.

^ The character typically above the 6 on your keyboard. It was originally a circumflex. In computer language it became the symbol that was written to represent the Control (Ctrl) key. It's also called the "hat."

This character is a tilde. It tells you how to pronounce the "n" in señor. According to William Safire, it's a Spanish word from the Latin term for a tiny diacritical mark used to change the phonetic value of a letter.

PUBLISHED BY TELECOM LIBRARY

Telecom Library publishes books and magazines and organizes trade conferences on telephones, telecommunications, and voice processing. It also distributes the books of other publishers, making it the "central source" for all the above materials. Call or write for your FREE catalog.

- Newton's TELECOM Dictionary
- T-1 Networking
- Negotiating Telecommunications Contracts
- Buying Short Haul Microwave
- The Perfect Proposal
- The Inbound Telephone Call Center
- Student Communications Services
- The Complete Traffic Engineering Handbook
- SONET: Planning, Installing &Maintaining Broadband Networks
- Speech Recognition The Complete Practical Reference Guide
- The Guide to Frame Relay and Fast Packet Networking
- Frames, Packets and Cells in Broadband Networking
- Which Phone System Should I Buy?
- Customer Service Over the Phone
- The Dictionary of Sales and Marketing Technology Terms
- Guide to Local and Long Distance Billing Practices
- SCSA: TheCompleteReference Guide

QUANTITY PURCHASES

If you wish to purchase this book, or any others, in quantity, please contact:
Christine Fullam-Kern, Manager
Telecom Library Inc.
12 West 21 Street
New York, NY 10010
1-800-LIBRARY or 212-691-8215
Facsimile orders: 212-691-1191

NOTES

NOTES

NOTES

NOTES

NOTES